Birkhäuser

Frontiers in Mathematics

This series is designed to be a repository for up-to-date research results which have been prepared for a wider audience. Graduates and postgraduates as well as scientists will benefit from the latest developments at the research frontiers in mathematics and at the "frontiers" between mathematics and other fields like computer science, physics, biology, economics, finance, etc. All volumes are online available at SpringerLink.

Arif Salimov

Applications of Holomorphic Functions in Geometry

Birkhäuser

Arif Salimov
Department of Algebra and Geometry
Baku State University
Baku, Azerbaijan

ISSN 1660-8046 ISSN 1660-8054 (electronic)
Frontiers in Mathematics
ISBN 978-981-99-1298-8 ISBN 978-981-99-1296-4 (eBook)
https://doi.org/10.1007/978-981-99-1296-4

This book is published under the imprint Birkhäuser, www.birkhauser-science.com by the registered company Springer Nature Singapore Pte Ltd.
The registered company address is: 152 Beach Road, #21-01/04 Gateway East, Singapore 189721, Singapore

Preface

This book is intended to provide a systematic introduction to the theory of holomorphic manifolds. The book furnishes detailed information on holomorphic functions in algebras and discusses some of the areas in geometry with applications. Its goal is to expound the recent developments in applications of holomorphic functions in the theory of hypercomplex and anti-Hermitian manifolds as well as in the geometry of bundles.

In spite of the geometric applications of holomorphic functions that are mainstream in the investigation of differential geometry, holomorphic manifolds and their recent applications are not so well known yet. The theory of holomorphic manifolds is more than 60 years old. The initial notion of a holomorphic manifold over algebras appeared in the 1960s in a series of papers of A.P. Shirokov and culminated in the book [1]. Since then, the subject has enjoyed a rapid development. Holomorphic manifolds relate with such topics as anti-Hermitian metrics and lifting of differential-geometrical objects to vector bundle.

The book is organized into three chapters. Chapter 1 presents the fundamental notions and some theorems concerning holomorphic functions and holomorphic manifolds, which are needed for the later applications. Section 1.1 gives the basic definitions and theorems on hypercomplex algebras. Section 1.2 is devoted to the study of holomorphic functions in algebra. Section 1.3 introduces the hypercomplex structures on manifolds. Section 1.4 treats manifolds with integrable regular hypercomplex structures. We show that such a manifold is a realization of a holomorphic manifold over algebra. Section 1.5 is devoted to the study of pure tensor fields. Here, we find the explicit expression of the pure tensor field with respect to the regular hypercomplex structures, which shows that the pure tensor fields on real manifolds are a realization of hypercomplex tensors. Section 1.6 discusses holomorphic hypercomplex tensor fields, and by using the Tachibana operator, we give the condition of holomorphic tensors in real coordinates. In Sect. 1.7, we consider pure connections which are realizations of the hypercomplex connections. Section 1.8 is devoted to the study of pure hypercomplex torsion tensors. In Sect. 1.9, we give a realization of holomorphic hypercomplex connections by using the pure curvature tensors. Finally, in Sect. 1.10, we consider some properties of pure curvature tensors. The main theorem of this section is that the curvature tensor of a holomorphic connection is holomorphic.

In Chap. 2, we study the pseudo-Riemannian metric on holomorphic manifolds. In Sect. 2.1, we give the condition for a hypercomplex anti-Hermitian metric to be holomorphic. In Sect. 2.2, we discuss complex Norden manifolds. We define the twin Norden metric. The main theorem of this section is that the Levi-Civita connection of Kähler-Norden metric coincides with the Levi-Civita connection of twin Norden metric. In Sect. 2.3, we consider Norden-Hessian structures. We give the condition for a Norden-Hessian manifold to be Kähler. Section 2.4 is devoted to the analysis of twin Norden metric connections with torsion. In Sects. 2.5-2.10, we focus our attention to pseudo-Riemannian 4-manifolds of neutral signature. The main purpose of these sections is to study complex Norden metrics on 4-dimensional Walker manifolds.

In the first part of Chap. 3, we focus on lifts from a manifold to its tensor bundle. Some introductory material concerning the tensor bundle is provided in Sect. 3.1. Section 3.2 is devoted to the study of the complete lifts of (1, 1)-tensor fields along cross-sections in the tensor bundle. In Sect. 3.3, we study holomorphic cross-sections of tensor bundles. In the second part of Chap. 3, we concentrate our attention to lifts from a manifold to its tangent bundles of orders 1 and 2 by using the realization of holomorphic manifolds. The main purpose of Sects. 3.4-3.9 is to study the differential-geometrical objects on the tangent bundle of order 1 corresponding to dual-holomorphic objects of the dual-holomorphic manifold. As a result of this approach, we find a new class of lifts, that is, deformed complete lifts of functions, vector fields, forms, tensor fields and linear connections in the tangent bundle of order 1. Section 3.10 is devoted to the study of holomorphic metrics in the tangent bundle of order 2 (that is, in the bundle of 2-jets) by using the Tachibana operator. By using the algebraic approach, the problem of deformed lifts of functions, vector fields and 1-forms is solved in Sects. 3.11-3.12. In Sect. 3.13, we investigate the complete lift of the almost complex structure to cotangent bundle and prove that it is a transfer by a symplectic isomorphism of complete lift to tangent bundle if the symplectic manifold with almost complex structure is an almost holomorphic A-manifold. Finally, in Sect. 3.14, we transfer via the differential of the musical isomorphism defined by pseudo-Riemannian metrics the complete lifts of vector fields and almost complex structures from the tangent bundle to the cotangent bundle.

The author believes that geometric applications of holomorphic functions are a very fruitful research domain and provides many new problems in the study of modern differential geometry. Researchers in the geometry, algebra, topology and physics communities may find the book useful as a self-study guide.

I warmly thank Simona-Luiza Druţă-Romaniuc from the Department of Mathematics and Informatics, Gheorghe Asachi Technical University of Iasi, Romania, for reading the manuscript and making useful corrections in it.

Baku, Azerbaijan

Arif Salimov

Reference

1. Vishnevskii, V.V., Shirokov, A.P., Shurygin, V.V.: Spaces over algebras. Kazanskii Gosudarstvennii Universitet, Kazan (1985)

Contents

About the Author

Arif Salimov is Full Professor and Head of the Department Algebra and Geometry, Faculty of Mechanics and Mathematics, Baku State University. An Azerbaijani/Soviet mathematician, honoured scientist of Azerbaijan, he is known for his research in differential geometry. He earned his B.Sc. degree from Baku State University, Azerbaijan, in 1978, a Ph.D. and Doctor of Sciences (Habilitation) degrees in geometry from Kazan State University, Russia, in 1984 and 1998, respectively. His advisor was Vladimir Vishnevskii. He is an author/co-author of more than 100 research papers. His primary areas of research are theory of lifts in tensor bundles, geometrical applications of tensor operators, special Riemannian manifolds, indefinite metrics and general geometric structures on manifolds (almost complex, almost product, hypercomplex, Norden structures, etc.).

Holomorphic Manifolds Over Algebras

1

In this chapter, we give the fundamental notions and some theorems concerning holomorphic functions and holomorphic manifolds which will be needed for the later applications. In Sect. 1.1, we give the basic definitions and theorems on hypercomplex algebras. Section 1.2 is devoted to the study of holomorphic functions in algebra. In Sect. 1.3, we introduce the hypercomplex structures on manifolds. Section 1.4 treats manifolds with integrable regular hypercomplex structures. We show that such a manifold is a realization of a holomorphic manifold over algebra. Section 1.5 is devoted to the study of pure tensor fields. We find the explicit expression of the pure tensor field with respect to the regular hypercomplex structures, and we show that the pure tensor fields on real manifolds are a realization of hypercomplex tensors. In Sect. 1.6, we discuss holomorphic hypercomplex tensor fields, and using the Tachibana operator, we give the condition of holomorphic tensors in real coordinates. In Sect. 1.7, we consider pure connections which are realizations of the hypercomplex connections. Section 1.8 is devoted to the study of pure hypercomplex torsion tensors. In Sect. 1.9, we give a realization of holomorphic hypercomplex connections by using the pure curvature tensors. In the last Sect. 1.10, we consider some properties of pure curvature tensors. The main theorem of this section is that the curvature tensor of a holomorphic connection is holomorphic. Finally, we also consider a holomorphic manifold of hypercomplex dimension 1, and we show that the hypercomplex connection on such a manifold is holomorphic if and only if the real manifold is locally flat.

1.1 Some Basic Concepts of Algebra

We consider an m-dimensional associative algebra $\mathfrak{A}_{\mathrm{m}}$ over the field of real numbers $\mathbb{R}$ (hypercomplex algebra) with basis $\{e_\alpha\}$, $\alpha = 1, \ldots, m$ and structure constants $C_{\alpha\beta}^{\gamma}$:

A. Salimov, *Applications of Holomorphic Functions in Geometry*,
Frontiers in Mathematics, https://doi.org/10.1007/978-981-99-1296-4_1

$$e_\alpha e_\beta = C^\gamma_{\alpha\beta} e_\gamma.$$

We note that $C^\gamma_{\alpha\beta}$ are components of the tensor of type (1,2) in the vector space of $\mathfrak{A}_\mathrm{m}$. In this work, we suppose $\mathfrak{A}_\mathrm{m}$ is an algebra with the unit $e_1 = 1$.

We introduce the matrices

$$C_\alpha = (C^\gamma_{\alpha\beta}), \quad \widetilde{C}_\alpha = (C^\gamma_{\beta\alpha}), \quad \alpha, \beta, \gamma = 1, ..., m, \tag{1.1}$$

where γ denotes rows and β denotes columns of matrices C_α and $\widetilde{C}_\alpha$. Then the associativity condition

$$(e_\alpha e_\beta)e_\gamma = e_\alpha(e_\beta e_\gamma) \ \Leftrightarrow \ C^\sigma_{\alpha\beta} e_\sigma e_\gamma = e_\alpha C^\sigma_{\beta\gamma} e_\sigma \ \Leftrightarrow \ C^\sigma_{\alpha\beta} C^\tau_{\sigma\gamma} = C^\sigma_{\beta\gamma} C^\tau_{\alpha\sigma}$$

can be written in one of the following three equivalent forms:

$$C_\alpha C_\beta = C^\sigma_{\alpha\beta} C_\sigma, \tag{1.2}$$

$$\widetilde{C}^T_\alpha \widetilde{C}^T_\beta = C^\gamma_{\alpha\beta} \widetilde{C}^T_\gamma, \tag{1.3}$$

$$C_\alpha \widetilde{C}_\beta = \widetilde{C}_\beta C_\alpha, \tag{1.4}$$

where $\widetilde{C}^T_\alpha$ is the tranpose of $\widetilde{C}_\alpha$. From $1 = \varepsilon^\sigma e_\sigma$ follows that

$$e_\alpha = \varepsilon^\sigma e_\sigma e_\alpha = \varepsilon^\sigma C^\tau_{\sigma\alpha} e_\tau, \quad e_\alpha = e_\alpha \varepsilon^\sigma e_\sigma = \varepsilon^\sigma C^\tau_{\alpha\sigma} e_\tau \ \ (e_\alpha = \delta^\tau_\alpha e_\tau)$$

or

$$\varepsilon^\sigma C_\sigma = \varepsilon^\sigma \widetilde{C}_\sigma = I \quad (\varepsilon^\sigma C^\tau_{\sigma\alpha} = \varepsilon^\sigma C^\tau_{\alpha\sigma} = \delta^\tau_\alpha), \tag{1.5}$$

where δ^τ_α is the Kronecker delta.

Let $C(\mathfrak{A})$ and $\widetilde{C}^T(\mathfrak{A})$ be the algebras of matrices of types C_α and $\widetilde{C}^T_\alpha$, respectively (see (1.2) and (1.3)). A mapping $\rho_1 : \mathfrak{A}_m \to C(\mathfrak{A})$:

$$\mathfrak{A}_m \ni a = a^\sigma e_\sigma \to a^\sigma C_\sigma = C(a) \in C(\mathfrak{A}), \quad a^1, ..., a^m \in \mathbb{R}$$

is called a *regular representation of type I*. The representation ρ_1 is obviously isomorphic. By similar devices, we introduce a regular representation $\rho_2 : \mathfrak{A}_m \to \widetilde{C}^T(\mathfrak{A})$ of type II:

$$\mathfrak{A}_m \ni a = a^\sigma e_\sigma \to a^\sigma \widetilde{C}^T_\sigma = \widetilde{C}^T(a) \in \widetilde{C}^T(\mathfrak{A}), \quad a^1, ..., a^m \in \mathbb{R}$$

which is also isomorphic. In this work, the representations appearing in the discussion will be supposed to be of type I. Notice that a regular representation of type I is usually called simply a regular representation.

From (1.4), we see that all $A = a^\sigma \widetilde{C}_\sigma, a^1, ..., a^m \in \mathbb{R}$ belongs to the commutator of the algebra $C(\mathfrak{A})$. Conversely, using (1.5), we easily see that if AC = CA for any $C \in C(\mathfrak{A})$, then $A = a^\alpha \widetilde{C}_\alpha$. In fact, we put $C = C_\alpha, \alpha = 1, ..., m$, then we have $A^\sigma_\gamma C^\gamma_{\alpha\beta} = C^\sigma_{\alpha\gamma} A^\gamma_\beta$. Contracting this equation with ε^β, we find

$$A^\sigma_\gamma C^\gamma_{\alpha\beta}\varepsilon^\beta = C^\sigma_{\alpha\gamma} A^\gamma_\beta \varepsilon^\beta,$$
$$A^\sigma_\gamma \delta^\gamma_\alpha = C^\sigma_{\alpha\gamma} a^\gamma,$$
$$A = (A^\sigma_\gamma) = a^\gamma (C^\sigma_{\alpha\gamma}) = a^\gamma \widetilde{C}_\gamma,$$

where $a^\gamma = A^\gamma_\beta \varepsilon^\beta$. Thus, we have

Theorem 1.1 *Let A be a square matrix of order m. Then* $AC_\alpha = C_\alpha A$ *if and only if* $A = a^\gamma \widetilde{C}_\gamma$.

By similar devices, we have

Theorem 1.2 *Let A be a square matrix of order m. Then* $A\widetilde{C}^T_\alpha = \widetilde{C}^T_\alpha A$ *if and only if* $A = a^\gamma \widetilde{C}^T_\gamma$.

Now, we restrict ourselves to the consideration of commutative hypercomplex algebras. In this and next sections, we always assume that the $\mathfrak{A}_m$ is a commutative hypercomplex algebra. The commutativity condition $e_\alpha e_\beta = e_\beta e_\alpha$ can be written in the following form:

$$C^\gamma_{\alpha\beta} = C^\gamma_{\beta\alpha} \quad (C_\alpha = \widetilde{C}_\alpha) \tag{1.6}$$

Using (1.6), from (1.4), we have

$$C^\gamma_{\alpha\sigma} C^\sigma_{\beta\delta} = C^\gamma_{\beta\sigma} C^\sigma_{\alpha\delta} \quad (C_\alpha C_\beta = C_\beta C_\alpha). \tag{1.7}$$

Since $C_\alpha = \widetilde{C}_\alpha$, then we have $\widetilde{C}^T(\mathfrak{A}) = C^T(\mathfrak{A})$. Therefore, $C^T(\mathfrak{A})$ is called a *transpose regular representations* of commutative algebras. From Theorems 1.1 and 1.2, we have.

Theorem 1.3 *Let* $\mathfrak{A}_m$ *be a commutative hypercomplex algebra. Then* $C(\mathfrak{A})$ *and* $C^T(\mathfrak{A})$ *are maximal commutative subalgebras of the algebra of square matrices of order m.*

Among commutative algebras, a special role is played by *Frobenius algebras* for which there exist constants λ_γ such that

$$\varphi_{\alpha\beta} = C^\gamma_{\alpha\beta}\lambda_\gamma \tag{1.8}$$

form a nonsingular symmetric matrix and can be taken as components of the metric tensor of the Frobenius metric. Then, along with the basis $\{e_\alpha\}$, we can introduce the dual basis $\{e^\alpha\}$, where $e^\alpha = \varphi^{\alpha\beta} e_\beta$. For the Frobenius metric, we have the relations

$$\varphi_{\alpha\sigma} C^\sigma_{\gamma\beta} = \varphi_{\beta\sigma} C^\sigma_{\gamma\alpha}, \ \varphi^{\alpha\sigma} C^\beta_{\gamma\sigma} = \varphi^{\beta\sigma} C^\alpha_{\gamma\sigma}, \ e^\alpha e_\beta = C^\alpha_{\beta\gamma} e^\gamma, \ e^\alpha e^\beta = \varphi^{\alpha\sigma} C^\beta_{\gamma\sigma} e^\gamma, \varphi_{\alpha\beta}\varepsilon^\beta = \lambda_\alpha, \tag{1.9}$$

where ε^β are components of $e_1 = 1$.

1.2 Holomorphic Functions

Let $z = x^\alpha e_\alpha$ be an algebraic variable in $\mathfrak{A}_m$, where $x^\alpha (\alpha = 1, ..., m)$ are real variables. We introduce an algebraic function $w = w(z) \in \mathfrak{A}_m$ of variable z in the following form:

$$w = y^\beta(x) e_\beta,$$

where $y^\beta(x) = y^\beta(x^1, ..., x^m)$, $\beta = 1, ..., m$ are real-valued C^∞-functions. Let $\mathrm{d}z = \mathrm{d}x^\alpha e_\alpha$ and $\mathrm{d}w = \mathrm{d}y^\alpha e_\alpha$ be the differentials of z and w, respectively. We shall say that the function $w = w(z)$ is a *holomorphic function* if there exists a functions $w'(z)$ such that

$$\mathrm{d}w = w'(z)\mathrm{d}z \tag{1.10}$$

We shall call $w'(z)$ the *derivative* of $w = w(z)$.

Theorem 1.4 [28, 71, 78, p.87] *The algebraic function $w = w(z)$ is $\mathfrak{A}$-holomorphic if and only if the Scheffers conditions (the generalized Cauchy-Riemann conditions) hold:*

$$C_\alpha D = \mathrm{D}C_\alpha \tag{1.11}$$

where $D = \left(\frac{\partial y^\alpha}{\partial x^\beta}\right)$ is the Jacobian matrix of $y^\alpha(x)$.

Proof Let $w = w(z)$ be a holomorphic function. We put $w'(z) = \tilde{w}^\alpha e_\alpha$. Then from (1.10), we have

$$\mathrm{d}w = \mathrm{d}y^\alpha e_\alpha = \frac{\partial y^\alpha}{\partial x^\beta}\mathrm{d}x^\beta e_\alpha = \tilde{w}^\alpha e_\alpha \mathrm{d}x^\beta e_\beta = \tilde{w}^\alpha \mathrm{d}x^\beta C^\gamma_{\alpha\beta} e_\gamma,$$

From here, we obtain

$$\frac{\partial y^\gamma}{\partial x^\beta} = \tilde{w}^\alpha C^\gamma_{\alpha\beta} \tag{1.12}$$

Thus, the hypercomplex function $w = w(z)$ is a holomorphic function if and only if the Jakobian matrix $\left(\frac{\partial y^\alpha}{\partial x^\beta}\right)$ has the form (1.12). Contracting (1.12) with ε^β $(1 = \varepsilon^\beta e_\beta)$ and using (1.5), we find

$$\tilde{w}^\gamma = \varepsilon^\beta \frac{\partial y^\gamma}{\partial x^\beta},$$

i.e.

$$w'(z) = \varepsilon^\beta \frac{\partial y^\gamma}{\partial x^\beta} e_\gamma.$$

Now applying Theorem 1.1 to (1.12), we see that the condition (1.12) is equivalent to the Scheffers conditions. Thus, the theorem is proved.

The concept of the holomorphic hypercomplex functions can be immediately extended to the case of several algebraic variables. Let

$$z^v = x^{(v-1)m+\beta} e_\beta \ (v = 1, ..., r)$$

be an r variables in $\mathfrak{A}_m$. In fact the functions $w^u(z^1, ..., z^r) = y^{(u-1)m+\alpha}(x^1, ..., x^{rm})e_\alpha$, $u = 1, ..., r$ are a holomorphic functions of the variables $z^1, ..., z^r$ if and only if

$$C_\alpha D_{u,v} = D_{u,v} C_\alpha \tag{1.13}$$

for any u and v, where

$$D_{u,v} = \left(\frac{\partial y^{(u-1)m+\alpha}}{\partial x^{(v-1)m+\beta}}\right), \quad u, v = 1, ..., r$$

and (see proof of Theorem 1.4)

$$\frac{\partial w^u}{\partial z^v} = \varepsilon^\beta \frac{\partial y^{(u-1)m+\alpha}}{\partial x^{(v-1)m+\beta}} e_\alpha \tag{1.14}$$

Remark 1.1 From (1.14), it follows that the Jacobian matrix of functions $\varepsilon^\beta \frac{\partial y^{(u-1)m+\alpha}}{\partial x^{(v-1)m+\beta}}$ has components of the form $D' = \left(\varepsilon^\gamma \frac{\partial^2 y^{(u-1)m+\alpha}}{\partial x^{(w-1)m+\gamma} \partial x^{(v-1)m+\beta}}\right) = \varepsilon^\gamma \frac{\partial D_{u,v}}{\partial x^{(w-1)m+\gamma}}$, and therefore, it also satisfies the Scheffers conditions (1.13), i.e. exists the successive derivatives $\frac{\partial^2 w^u}{\partial z^v \partial z^t}, \frac{\partial^3 w^u}{\partial z^v \partial z^t \partial z^l}, ...$ of $w^u = w^u(z^1, ..., z^r)$.

Example 1.1 We note that, in particular, if $\mathfrak{A}_2 = \mathbb{C}$, where $\mathbb{C}$ is the complex algebra, then the Scheffers conditions reduce to the Cauchy-Riemann conditions. In fact, by virtue of

$$C_1 = \begin{pmatrix} C_{11}^1 & C_{12}^1 \\ C_{11}^2 & C_{12}^2 \end{pmatrix} = \begin{pmatrix} 1 & 0 \\ 0 & 1 \end{pmatrix}, \quad C_2 = \begin{pmatrix} C_{21}^1 & C_{22}^1 \\ C_{21}^2 & C_{22}^2 \end{pmatrix} = \begin{pmatrix} 0 & -1 \\ 1 & 0 \end{pmatrix}$$

from (1.11), we have

$$\frac{\partial y^1}{\partial x^1} = \frac{\partial y^2}{\partial x^2}, \quad \frac{\partial y^2}{\partial x^1} = -\frac{\partial y^1}{\partial x^2},$$

where $z = x^1 + ix^2,\ w = y^1(x^1, x^2) + iy^2(x^1, x^2),\ i^2 = -1.$

Example 1.2. Let $\mathfrak{A}_4 = B(1, i_1, i_2, e)$ be a bicomplex algebra with canonical basis $\{1, i_1, i_2, e\}$, $i_1^2 = i_2^2 = -1$, $i_1 i_2 = i_2 i_1 = e$, $e^2 = 1$. The algebra of bicomplex numbers is the first nontrivial complex Clifford algebra (and the only commutative one) and has recently been applied to quantum mechanics and fractal theory. Using the Scheffers conditions $C_\alpha D = DC_\alpha$, where $D = \left(\frac{\partial f^\alpha}{\partial x^\beta}\right)$ is the Jacobian matrix of bicomplex function

$$F = f^1(x^1, x^2, x^3, x^4) + i_1 f^2(x^1, x^2, x^3, x^4) + i_2 f^3(x^1, x^2, x^3, x^4) + e f^4(x^1, x^2, x^3, x^4)$$

and

$$C_1 = \begin{pmatrix} 1&0&0&0 \\ 0&1&0&0 \\ 0&0&1&0 \\ 0&0&0&1 \end{pmatrix}, \quad C_2 = \begin{pmatrix} 0&-1&0&0 \\ 1&0&0&0 \\ 0&0&0&-1 \\ 0&0&1&0 \end{pmatrix},$$

$$C_3 = \begin{pmatrix} 0&0&-1&0 \\ 0&0&0&-1 \\ 1&0&0&0 \\ 0&1&0&0 \end{pmatrix}, \quad C_4 = \begin{pmatrix} 0&0&0&1 \\ 0&0&-1&0 \\ 0&-1&0&0 \\ 1&0&0&0 \end{pmatrix},$$

we have: the bicomplex function F is bicomplex-holomorphic if and only if the function F satisfies the following Scheffers conditions:

$$\frac{\partial f^1}{\partial x^1} = \frac{\partial f^2}{\partial x^2} = \frac{\partial f^3}{\partial x^3} = \frac{\partial f^4}{\partial x^4}, \quad \frac{\partial f^1}{\partial x^2} = \frac{\partial f^3}{\partial x^4} = -\frac{\partial f^2}{\partial x^1} = -\frac{\partial f^4}{\partial x^3},$$

$$\frac{\partial f^1}{\partial x^3} = \frac{\partial f^2}{\partial x^4} = -\frac{\partial f^3}{\partial x^1} = -\frac{\partial f^4}{\partial x^2}, \quad \frac{\partial f^1}{\partial x^4} = \frac{\partial f^4}{\partial x^1} = -\frac{\partial f^2}{\partial x^3} = -\frac{\partial f^3}{\partial x^2}$$

We note that the above Scheffers conditions are equivalent to the following bicomplex Cauchy-Riemann equations:

$$\frac{\partial \psi_1}{\partial q_1} = \frac{\partial \psi_2}{\partial q_2}, \frac{\partial \psi_1}{\partial q_2} = -\frac{\partial \psi_2}{\partial q_1},$$

where $F = \psi_1(q_1, q_2) + i_2\psi_2(q_1, q_2)$, $\psi_1(q_1, q_2) = f^1 + i_1 f^2$, $\psi_2(q_1, q_2) = f^3 + i_1 f^4$, $q_1 = x^1 + i_1 x^2$, $q_2 = x^3 + i_1 x^4$.

1.3 Algebraic Π-Structures on Manifolds

If a collection of (1,1)-tensor (affinor) fields $\underset{1}{\varphi}, \underset{2}{\varphi}, ..., \underset{m}{\varphi}$ is given on a smooth manifold M of dimension n, then one says that a *polyaffinor structure* (or *Π-structure*) is given on M:

$$\Pi = \left\{\underset{\alpha}{\varphi}{}^i_j\right\}, \quad \alpha = 1, ..., m; \quad i, j = 1, ..., n.$$

If there exists an *adapted frame* $\{X_i\}$, $i = 1, ..., n$ such that the each structure affinors $\underset{\alpha}{\varphi}$ of Π-structure has constant components $\underset{\alpha}{\varphi}{}^i_j = const$ with respect to this frame, then the Π-structure is called a *rigid* structure [36]. In general, the frame $\{X_i\}$ is not a natural frame. If the adapted frame $\{X_i\}$ is a natural frame, i.e. $\{X_i\} = \{\partial_i\}$, $\partial_i \underset{\alpha}{\varphi}{}^i_j = 0$, then the Π-structure is said to be *integrable*. It is clear that the integrable Π-structure is rigid, and the contrary statement is true only under some additional conditions on Π-structure. For example, if $\Pi = \varphi$, i.e. if the Π-structure contains only one affinor φ, and if there exists a torsion-free connection ∇ on M preserving the rigid φ-structure, then the φ-structure is integrable [73]. It is well known that for simplest rigid φ-structures (almost complex and almost paracomplex structures, etc.), the integrability is equivalent to the vanishing of the Nijenhuis tensor.

Definition 1.1 Let ∇ be a linear connection on M. ∇ is called Π-connection with respect to the Π-structure, if $\nabla\varphi = 0$ for any $\varphi \in \Pi$.

Definition 1.2 Π-structure on M is called almost integrable, if there exists a torsion-free Π-connection.

We note that for some simplest Π-structures (regular Π-structure [27]), the concepts of integrability and almost integrability are equivalent.

Let $\mathfrak{A}_m$ be a commutative hypercomplex algebra with the unit $e_1 = 1$. An *almost hypercomplex structure* on M is a polyaffinor Π-structure such that

$$\underset{\alpha}{\varphi} \circ \underset{\beta}{\varphi} = C^{\gamma}_{\alpha\beta} \underset{\gamma}{\varphi}, \tag{1.15}$$

i.e. if there exists an isomorphism $\mathfrak{A}_m \leftrightarrow \Pi$, where $\underset{1}{\varphi} = Id_M, \underset{2}{\varphi}, ...\underset{m}{\varphi}$ are structure affinors corresponding to the base elements $e_1 = 1, e_2, ..., e_m \in \mathfrak{A}_m$.

Definition 1.3 An almost hypercomplex structure on M is said to be an r-regular Π-structure (or for simplicity a regular Π-structure) if the matrices $\underset{\alpha}{\varphi} = \left(\underset{\alpha}{\varphi}{}^i_j\right)$, $\alpha = 1, ..., m$ of order $n \times n$ simultaneously reduced to the form

$$\underset{\alpha}{\varphi} = \left(\underset{\alpha}{\varphi}{}^i_j\right) == \begin{pmatrix} C_\alpha & 0 & \cdots 0 \\ 0 & C_\alpha & \cdots 0 \\ \cdots & \cdots & \cdots \\ 0 & 0 & \cdots C_\alpha \end{pmatrix}, \alpha = 1, ..., m; i, j = 1, ..., n \tag{1.16}$$

with respect to the adapted frame $\{X_i\}$, where $C_\alpha = \left(C^\gamma_{\alpha\beta}\right)$ is the regular representation of $\mathfrak{A}_r$ and r is a number of C_α—blocks.

From Definition 1.3 immediately follows that the regular Π-structures are rigid structures. In particular, for almost complex and paracomplex structures, the condition (1.16) immediately follows from (1.15), i.e. almost complex and paracomplex structures (see [10, 19]) on $M(\dim M = n = 2r)$ automatically are regular structures.

Let now $\Pi = \left\{\underset{\alpha}{\varphi}\right\}$, $\alpha = 1, ..., m$ be a regular Π-structure on M. Then from (1.16), we have

$$n = mr(n = \dim M, m = \dim \mathfrak{A}_m), \tag{1.17}$$

where r is a number of the blocks C_α. Thus, the condition (1.17) is a necessary condition for the existence of regular Π-structures on M_{mr}, and in this case, we have

$$i = (u-1)m + \alpha \quad (i = 1, ..., n; u = 1, ..., r; \alpha = 1, ..., m)$$

or

$$i = u\alpha, j = v\beta, k = w\gamma, ...$$

In other words, the structure affinors $\underset{\alpha}{\varphi}$ have the components

$$\underset{\sigma}{\varphi}{}^i_j = \underset{\sigma}{\varphi}{}^{u\alpha}_{v\beta} = \delta^u_v C^\alpha_{\sigma\beta} (\delta^u_v\text{-Kronecker delta}). \tag{1.18}$$

Let $X_{i'} = S^i_{i'} X_i (Det(S^i_{i'}) \neq 0)$ be a transformation of adapted frame $\{X_i\}$ with respect to the regular Π-structure. Then we have the following matrix relationship:

$$\begin{pmatrix} \varphi^{i'}_{j'} \\ \alpha \end{pmatrix} = \left(S^{i'}_{i}\right)\begin{pmatrix} \varphi^{i}_{j} \\ \alpha \end{pmatrix}\left(S^{j}_{j'}\right), \tag{1.19}$$

where $(S^{i'}_{i}) = (S^{i}_{i'})^{-1}$. If $\{X_{i'}\}$ is an adapted frame, then we call transformation $X_i \to X_{i'}$ the admissible transformation. Since $\begin{pmatrix} \varphi^{i'}_{j'} \\ \alpha \end{pmatrix} = \begin{pmatrix} \varphi^{i}_{j} \\ \alpha \end{pmatrix}$ for the adapted frames, we have from (1.19)

$$S \underset{\alpha}{\varphi} = \underset{\alpha}{\varphi} S, \tag{1.20}$$

where $S = \left(S^{j}_{j'}\right)$ and $\underset{\alpha}{\varphi} = \begin{pmatrix} \varphi^{i}_{j} \\ \alpha \end{pmatrix}$. Thus, we have

Theorem 1.5 *Let* $\Pi = \left\{\underset{\alpha}{\varphi}\right\}$ *be a regular* Π*-structure on* M_{mr}. *A transformation* $S : X_i \to X_{i'}$ *of adapted frames is admissible if and only if the condition* (1.20) *is true.*

Using Theorems 1.1 and 1.5, we see that the matrix S has the following special structure:

$$S^{i}_{i'} = \Delta^{u\sigma}_{u'} C^{\alpha}_{\sigma\alpha'} \quad (i = u\alpha, i' = u'\alpha') \tag{1.21}$$

By similar devices for inverse matrix, we have

$$S^{i'}_{i} = \Delta^{u'\sigma}_{u} C^{\alpha'}_{\sigma\alpha} \, (i = u\alpha, \;\; i' = u'\alpha') \tag{1.22}$$

We can introduce the following matrices in the algebra $\mathfrak{A}_m$:

$$\overset{*}{S} = \left(\overset{*}{S}{}^{u}_{u'}\right) = \left(\Delta^{u\sigma}_{u'} e_\sigma\right), \quad \overset{*}{S}{}^{-1} = \left(\overset{*}{S}{}^{u'}_{u}\right) = \left(\Delta^{u'\sigma}_{u} e_\sigma\right)$$

From here, we easily see that $\overset{*}{S}\overset{*}{S}{}^{-1} = \overset{*}{I}$, where

$$\overset{*}{I} = \begin{pmatrix} e_1 \, 0 & \cdots & 0 \\ 0 \;\, e_1 & \cdots & 0 \\ 0 \;\, 0 & \cdots & e_1 \end{pmatrix} = \begin{pmatrix} 1 \, 0 & \cdots & 0 \\ 0 \, 1 & \cdots & 0 \\ 0 \, 0 & \cdots & 1 \end{pmatrix} = \left(\delta^{u}_{v}\right).$$

In fact, from (1.21) and (1.22), we obtain

$$\begin{aligned} \overset{*}{S}{}^{u}_{u'} \overset{*}{S}{}^{u'}_{v} &= \Delta^{u\sigma}_{u'} e_\sigma \Delta^{u'\varepsilon}_{v} e_\varepsilon = \Delta^{u\sigma}_{u'} \Delta^{u'\varepsilon}_{v} C^{\gamma}_{\sigma\varepsilon} e_\gamma \\ &= \Delta^{u\sigma}_{u'} C^{\alpha}_{\sigma\alpha'} \Delta^{u'\varepsilon}_{v} C^{\alpha'}_{\varepsilon\beta} = S^{i}_{i'} S^{i'}_{j} = \delta^{i}_{j} = \delta^{u}_{v} C^{\alpha}_{1\beta} = \delta^{u}_{v} e_1 = \delta^{u}_{v}. \end{aligned}$$

For any vector field $\xi = \xi^i X_i = \xi^{u\alpha} X_{u\alpha}$, where $\{X_i\}$ is the adapted frame on M_{mr}, we can associate r coordinates $\overset{*}{\xi}{}^u$ $(u = 1, ..., r)$ in the algebra $\mathfrak{A}_m$:

$$\overset{*}{\xi}{}^u = \xi^{u\alpha} e_\alpha.$$

We easily see that, if $\xi^{i'} = S_i^{i'} \xi^i$, then $\overset{*}{\xi}{}^{u'} = \overset{*}{S}{}_u^{u'} \overset{*}{\xi}{}^u$. In fact, from (1.21), we obtain

$$\xi^{u'\alpha'} = S_{u\alpha}^{u'\alpha'} \xi^{u\alpha} = \Delta_u^{u'\sigma} C_{\sigma\alpha}^{\alpha'} \xi^{u\alpha}$$

or

$$\begin{aligned}\overset{*}{\xi}{}^{u'} &= \xi^{u'\alpha'} e_{\alpha'} = \Delta_u^{u'\sigma} C_{\sigma\alpha}^{\alpha'} \xi^{u\alpha} e_{\alpha'} \\ &= \Delta_u^{u'\sigma} e_\sigma \xi^{u\alpha} e_\alpha = \overset{*}{S}{}_u^{u'} \overset{*}{\xi}{}^u.\end{aligned}$$

Let now $\varphi \in \Pi$. Then $\varphi = a^\alpha \underset{\alpha}{\varphi}$. The action of operator φ for the vector field $\xi = \xi^i X_i$, i.e. the equation $\eta^i = \varphi_j^i \xi^j$, reduces to

$$\begin{aligned}\overset{*}{\eta}{}^u &= \eta^{u\alpha} e_\alpha = a^\sigma \underset{\sigma}{\varphi_j^i} \xi^j e_\alpha = a^\sigma \delta_v^u C_{\sigma\beta}^\alpha \xi^{v\beta} e_\alpha \\ &= a^\sigma \delta_v^u \xi^{v\beta} e_\sigma e_\beta = a^\sigma e_\sigma \xi^{u\beta} e_\beta = a \overset{*}{\xi}{}^u,\end{aligned}$$

where $i = u\alpha$, $j = v\beta$ and $a \in \mathfrak{A}_m$. Thus, we have

Theorem 1.6 *If Π is a regular structure on M_{mr}, then each tangent space $T_x(M_{mr})$, $x \in M_{mr}$ serves as a realization of the module $T_r(\mathfrak{A}_m)$ over algebra $\mathfrak{A}_m$.*

In particular, for the case, where $\varphi = \underset{\alpha}{\varphi}$, from $\overset{*}{\eta}{}^u = a \overset{*}{\xi}{}^u$, we have

$$\overset{*}{\eta}{}^u = e_\alpha \overset{*}{\xi}{}^u.$$

1.4 Integrable Structures and Holomorphic Manifolds

Let $\Pi = \left\{\underset{\alpha}{\varphi}\right\}$ be an integrable regular Π-structure on M_{mr}, and let $x^i = x^{u\alpha}$ and $x^{i'} = x^{u'\alpha'}$ be an adapted local coordinates in $U_x (x \in M_{mr})$. It is well known that the affinors $\underset{\alpha}{\varphi}$ have the constant form (1.18) with respect to the adapted frames $\left\{\frac{\partial}{\partial x^i}\right\}$ and $\left\{\frac{\partial}{\partial x^{i'}}\right\}$, and the admissible transformation has the form

$$\frac{\partial}{\partial x^{i'}} = S^i_{i'} \frac{\partial}{\partial x^i}$$

with $S^i_{i'} = \frac{\partial x^i}{\partial x^{i'}}$, $S^i_{i'} = \Delta^{u\sigma}_{u'} C^{\alpha}_{\sigma\alpha'}$, $S^{i'}_i = \Delta^{u'\sigma}_{u} C^{\alpha'}_{\sigma\alpha}$, $i = u\alpha$, $i' = u'\alpha'$ (see (1.21) and (1.22)). Then from Theorem 1.5, we have

$$\left(\frac{\partial x^{u\alpha}}{\partial x^{u'\alpha'}}\right) C_\sigma = C_\sigma \left(\frac{\partial x^{u\alpha}}{\partial x^{u'\alpha'}}\right)$$

for fixed u and u', i.e. by virtue of (1.13), the hypercomplex functions $z^{u'} = x^{u'\alpha'} e_{\alpha'}$ are the holomorphic functions of $z^u = x^{u\alpha} e_\alpha$. Using (1.14) and (1.22), we have

$$\frac{\partial z^{u'}}{\partial z^u} = \varepsilon^\alpha \frac{\partial x^{u'\alpha'}}{\partial x^{u\alpha}} e_{\alpha'} = \varepsilon^\alpha \Delta^{u'\sigma}_{u} C^{\alpha'}_{\alpha\sigma} e_{\alpha'} = \Delta^{u'\sigma}_{u} \delta^{\alpha'}_{\sigma} e_{\alpha'} = \Delta^{u'\sigma}_{u} e_\sigma = \overset{*}{S}{}^{u'}_{u}.$$

By similar devices, we have

$$\frac{\partial z^u}{\partial z^{u'}} = \overset{*}{S}{}^{u}_{u'}.$$

Thus in the intersection of any two coordinate charts with local adapted coordinates $x^{u\alpha}$ and $x^{u'\alpha'}$ of atlas on M_{mr}, the transition functions $z^{u'} = z^{u'}(z^u)$ $(z^u = x^{u\alpha} e_\alpha,\ z^{u'} = x^{u'\alpha'} e_{\alpha'})$ are $\mathfrak{A}$-holomorphic, i.e. a real C^∞-manifold M_{mr} of dimension mr also possesses the structure of a holomorphic $\mathfrak{A}$-manifold $X_r(\mathfrak{A})$ of dimension r. Thus, we have.

Theorem 1.7 *A realization of a holomorphic $\mathfrak{A}$-manifold $X_r(\mathfrak{A})$ is a real C^∞-manifold M_{mr} with integrable regular Π-structure.*

1.5 Pure Tensors

Let now M be a C^∞-manifold of finite dimension n. We denote by $\Im^r_s(M)$ the module over $F(M)$ of all C^∞-tensor fields of type (r, s) on M, where $F(M)$ is the algebra of C^∞ functions on M.

Definition 1.4 Let φ be an affinor field on M, i.e. $\varphi \in \Im^1_1(M)$. A tensor field t of type (r, s) is called pure tensor field with respect to φ if

$$t\left(\varphi X_1, X_2, ..., X_s, \overset{1}{\xi}, \overset{2}{\xi}, ..., \overset{r}{\xi}\right) = t\left(X_1, \varphi X_2, ..., X_s, \overset{1}{\xi}, \overset{2}{\xi}, ..., \overset{r}{\xi}\right)$$

$$\vdots$$

$$= t\left(X_1, X_2, ..., \varphi X_s, \overset{1}{\xi}, \overset{2}{\xi}, ..., \overset{r}{\xi}\right)$$

$$= t\left(X_1, X_2, ..., X_s, \varphi' \overset{1}{\xi}, \overset{2}{\xi}, ..., \overset{r}{\xi}\right)$$

$$= t\left(X_1, X_2, ..., X_s, \overset{1}{\xi}, \varphi' \overset{2}{\xi}, ..., \overset{r}{\xi}\right)$$

$$\vdots$$

$$= t\left(X_1, X_2, ..., X_s, \overset{1}{\xi}, \overset{2}{\xi}, ..., \varphi' \overset{r}{\xi}\right)$$

for any $X_1, X_2, ..., X_s \in \Im_0^1(M)$ and $\overset{1}{\xi}, \overset{2}{\xi}, ..., \overset{r}{\xi} \in \Im_1^0(M)$, where φ' is the adjoint operator of φ defined by $(\varphi'\xi)(X) = \xi(\varphi X)$.

Let $x^1, x^2, ..., x^n$ be a local coordinates in M. By setting $X_1 = \frac{\partial}{\partial x^{i_1}}, ..., X_s = \frac{\partial}{\partial x^{i_s}}$ and $\overset{1}{\xi} = dx^{j_1}, ..., \overset{r}{\xi} = dx^{j_r}$, we see that the condition of pure tensor fields may be expressed in terms of the components φ_j^i and $t_{i_1...i_s}^{j_1...j_r}$ as follows:

$$t_{mi_2...i_s}^{j_1...j_r} \varphi_{i_1}^m = t_{i_1 m...i_s}^{j_1...j_r} \varphi_{i_2}^m = ... = t_{i_1 i_2...m}^{j_1...j_r} \varphi_{i_s}^m = t_{i_1...i_s}^{m j_2...j_r} \varphi_m^{j_1} = t_{i_1...i_s}^{j_1 m...j_r} \varphi_m^{j_2} = ... = t_{i_1...i_s}^{j_1 j_2...m} \varphi_m^{j_r} \tag{1.23}$$

We consider for convenience sake the vector, covector and scalar fields as pure tensor fields. Different problems concerning pure tensor fields have been studied by many authors (see, for example, [6, 7, 13, 17, 18, 20, 22, 23, 45, 50, 54–56, 60, 65–67, 74, 75, 77, 80, 86]).

In particular, let now $t \in \Im_1^1(M)$ be a pure tensor field of type (1.1). Then the purity condition may be written as:

$$\varphi(tX) = t(\varphi X).$$

Thus, if $t \in \Im_1^1(M)$ and $\varphi \in \Im_1^1(M)$ satisfy the commutativity condition

$$\varphi \circ t = t \circ \varphi, \tag{1.24}$$

where $(\varphi \circ t)X = \left(\varphi \overset{C}{\otimes} t\right)X = \varphi(tX)$ ($\overset{C}{\otimes}$ is a tensor product with a contraction C), then t is pure with respect to φ, and conversely, φ is also pure with respect to t.

From (1.24), it follows easily that φ itself and the unit affinor field I are examples of the pure tensor field. Also, from (1.24), we have: if φ is a regular affinor field, i.e. $\det(\varphi_j^i) \neq 0$, then the affinor field φ^{-1} whose components are given by the elements of the inverse matrix of φ is also pure.

In particular, being applied to a (1, 1)—tensor field t, the purity condition with respect to the regular Π-structure $\Pi = \left\{\underset{\gamma}{\varphi}\right\}$ means that in the local coordinates, the following conditions should hold:

$$t_j^m \underset{\gamma}{\varphi}{}_m^i = t_m^i \underset{\gamma}{\varphi}{}_j^m \tag{1.25}$$

We set $i = u\alpha$, $j = v\beta$, $m = w\sigma$ and $\underset{\gamma}{\varphi}{}_j^i = \delta_v^u C_{\gamma\beta}^\alpha$. Then, from (1.25), we have

$$t_{v\beta}^{w\sigma} \delta_w^u C_{\gamma\sigma}^\alpha = t_{w\sigma}^{u\alpha} \delta_v^w C_{\gamma\beta}^\sigma \Leftrightarrow t_{v\beta}^{u\sigma} C_{\gamma\sigma}^\alpha = t_{v\sigma}^{u\alpha} C_{\gamma\beta}^\sigma. \tag{1.26}$$

Using contraction with ε^β and (1.5), from (1.26), we have

$$t_{v\beta}^{u\sigma} \varepsilon^\beta C_{\gamma\sigma}^\alpha = t_{v\sigma}^{u\alpha} C_{\gamma\beta}^\sigma \varepsilon^\beta = t_{v\sigma}^{u\alpha} \delta_\gamma^\sigma = t_{v\gamma}^{u\alpha}$$

or

$$t_j^i = t_{v\beta}^{u\alpha} = \overset{\sigma}{\Im}{}_v^u C_{\sigma\beta}^\alpha, \tag{1.27}$$

where $\overset{\sigma}{\Im}{}_v^u = t_{v\gamma}^{u\sigma} \varepsilon^\gamma$. Thus, a pure tensor field $t \in \Im_1^1(M)$ has the form (1.27).

Conversely, from (1.27), it follows that the tensor field t of type (1.1) is pure if $\overset{\sigma}{\Im}$ are arbitrary functions. In fact, substituting (1.27) into (1.25), we find

$$\overset{\varepsilon}{\Im}{}_v^w C_{\varepsilon\beta}^\sigma \delta_w^u C_{\gamma\sigma}^\alpha = \overset{\varepsilon}{\Im}{}_w^u C_{\varepsilon\sigma}^\alpha \delta_v^w C_{\gamma\beta}^\sigma \Leftrightarrow \overset{\varepsilon}{\Im}{}_v^u \left(C_{\varepsilon\beta}^\sigma C_{\gamma\sigma}^\alpha - C_{\varepsilon\sigma}^\alpha C_{\gamma\beta}^\sigma \right) = 0. \tag{1.28}$$

Since $\mathfrak{A}_m$ is a commutative algebra $\left(C_{\varepsilon\beta}^\sigma C_{\gamma\sigma}^\alpha = C_{\varepsilon\sigma}^\alpha C_{\gamma\beta}^\sigma \Leftrightarrow C_\gamma C_\varepsilon = C_\varepsilon C_\gamma \right)$, we see that Eq. (1.28) is satisfied for arbitrary functions $\overset{\varepsilon}{\Im}{}_v^u$.

Thus, the tensor field t of type (0,2) is pure with respect to the regular Π-structure if and only if t has the form (1.27).

In the case $g \in \overset{*}{\Im}{}_2^0(M)$, by similar devices, we see that the tensor field g of type (0,2) is pure if and only if g has the form

$$g_{ij} = g_{u\alpha v\beta} = \Im_{uv\sigma} C_{\alpha\beta}^\sigma$$

for any functions $\Im_{uv\sigma} = g_{u\gamma v\sigma} \varepsilon^\gamma$ (for more details see Chap. 2).

In the case the situation is very difficult. The purity condition of G is given by

$$G^{mj} \underset{\gamma}{\varphi}{}_m^i = G^{im} \underset{\gamma}{\varphi}{}_j^m.$$

In a similar way, we have

$$G^{u\sigma v\beta}C^{\alpha}_{\gamma\sigma} = G^{u\alpha v\sigma}C^{\beta}_{\gamma\sigma}. \tag{1.29}$$

After contraction of (1.29) with any covector λ_α, we find

$$\varphi_{\gamma\sigma}G^{u\sigma v\beta} = \overset{\sigma}{\mathfrak{J}}{}^{uv}\, C^{\beta}_{\gamma\sigma}\left(\overset{\sigma}{\mathfrak{J}}{}^{uv} = G^{u\alpha v\sigma}\lambda_\alpha\right), \tag{1.30}$$

where $\varphi_{\gamma\sigma} = \lambda_\alpha C^{\alpha}_{\gamma\sigma}$. If $\mathrm{Det}(\varphi_{\gamma\sigma}) \neq 0$, i.e. if $\mathfrak{A}_m$ is a Frobenius algebra, then from (1.30), we have the following solution:

$$G^{ij} = G^{u\alpha v\beta} = \overset{\sigma}{\mathfrak{J}}{}^{uv}C^{\beta}_{\gamma\sigma}\varphi^{\gamma\alpha}. \tag{1.31}$$

Conversely, from (1.31) by virtue of (1.7), it follows that the tensor field $G \in \overset{*}{\mathfrak{J}}{}^2_0(M)$ is pure if $\overset{\sigma}{\mathfrak{J}}{}^{\text{uv}}$ are arbitrary functions. Thus, in the case when $G \in \overset{*}{\mathfrak{J}}{}^2_0(M)$, the algebra must be a Frobenius algebra.

In the general case, when $t \in \mathfrak{J}^r_s(M)$, in the space of $\mathfrak{A}_m$,we introduce the Kruchkovich tensors [28, 29]:

$$B^{\alpha}_{\beta_1\beta_2\ldots\beta_s} = C^{\alpha}_{\beta_1\sigma_1}C^{\sigma_1}_{\beta_2\sigma_2}\cdots C^{\sigma_{s-2}}_{\beta_{s-1}\beta_s}\,(s > 2),$$
$$B^{\alpha}_{\beta_1\beta_2} = C^{\alpha}_{\beta_1\beta_2},\ B^{\alpha}_{\beta} = \delta^{\alpha}_{\beta}.$$

If $\mathfrak{A}_m$ is the Frobenius algebra with metric $\varphi_{\alpha\beta}$, then we have

$$B^{\alpha_1\cdots\alpha_r}_{\beta_1\cdots\beta_s} = B^{\alpha_r}_{\beta_1\cdots\beta_s\lambda_1\cdots\lambda_{r-1}}\varphi^{\lambda_1\alpha_1}\cdots\varphi^{\lambda_{r-1}\alpha_{r-1}},$$
$$B^{\alpha_1\cdots\alpha_r} = B^{\alpha_r}_{\lambda_1\cdots\lambda_{r-1}}\varphi^{\lambda_1\alpha_1}\cdots\varphi^{\lambda_{r-1}\alpha_{r-1}},$$
$$B_{\beta_1\cdots\beta_s} = B^{\alpha}_{\beta_1,\cdots\beta_{s-1}}\varphi_{\alpha\beta_s}.$$

Now we state some properties of tensors $B^{\alpha_1\cdots\alpha_r}_{\beta_1\cdots\beta_s}$:

$(B_1)B^{\alpha_1\cdots\alpha_r}_{\beta_1\cdots\beta_s}$ is a symmetric tensor with respect to the indices $\alpha_1, \ldots, \alpha_r$ and $\beta_1, \ldots, \beta_s$,
$(B_2)C^{\lambda}_{\sigma\mu}B^{\sigma\alpha_1\cdots\alpha_r}_{\beta_1\cdots\beta_s} = B^{\lambda\alpha_1\cdots\alpha_r}_{\mu\beta_1\cdots\beta_s},\ C^{\sigma}_{\lambda\mu}B^{r_1\cdots r_s}_{\sigma\beta_1\cdots\beta_s} = B^{r_1\cdots r_s}_{\lambda\mu\beta_1\cdots\beta_s}$,
$(B_3)B^{\sigma\alpha_1\cdots\alpha_r}_{\beta_1\cdots\beta_s}\lambda_\sigma = B^{\alpha_1\cdots\alpha_r}_{\sigma\beta_1\cdots\beta_s}\varepsilon^{\sigma} = B^{\alpha_1\cdots\alpha_r}_{\beta_1\cdots\beta_s}$.

Proof of B_1,B_2,B_3 immediately follows from (1.5), (1.8) and (1.9).

In a similar way, the pure tensor field of type (r, s) has the following explicit expression

$$t^{i_1\cdots i_r}_{j_1\cdots j_s} = \overset{\sigma}{\mathfrak{J}}{}^{u_1\cdots u_r}_{v_1\cdots v_s}B^{\alpha_1\cdots\alpha_r}_{\sigma\beta_1\cdots\beta_s},$$
$$(i_a = u_a\alpha_a,\ j_b = v_b\beta_b,\ a = 1, \ldots, r, b = 1, \ldots, s) \tag{1.32}$$

where $\overset{\sigma}{\mathfrak{I}}{}^{u_1\cdots u_r}_{v_1\cdots v_s}$ are arbitrary functions in the adapted coordinate chart $U \subset M_{mr}$.

Using (1.32), we introduce a hypercomplex object in the Frobenius algebra $\mathfrak{A}_m$:

$$\overset{*}{t}{}^{u_1\ldots u_r}_{v_1\ldots v_s} = t^{i_1\ldots i_r}_{j_1\ldots j_s}\varepsilon^{\beta_1}\ldots\varepsilon^{\beta_s}\lambda_{\alpha_1}\ldots\lambda_{\alpha_{r-1}}e_{\alpha_r} = \overset{\sigma}{\mathfrak{I}}{}^{u_1\ldots u_r}_{v_1\ldots v_s}e_\sigma \tag{1.33}$$

We easily see that $\overset{*}{t}{}^{u_1\ldots u_r}_{v_1\ldots v_s}$ are components of hypercomplex tensor field of type (r, s), i.e.

$$\overset{*}{t}{}^{u'_1\cdots u'_r}_{v'_1\cdots v'_s} = \overset{*}{S}{}^{u'_1}_{u_1}\cdots\overset{*}{S}{}^{u'_r}_{u_r}\overset{*}{S}{}^{v_1}_{v'_1}\cdots\overset{*}{S}{}^{v_s}_{v'_s}\overset{*}{t}{}^{u_1\cdots u_r}_{v_1\cdots v_s}$$

For simplicity, we take $r = s = 1$. In fact, from (1.21) and (1.22), we have

$$t^{i'}_{j'} = t^{u'\alpha'}_{v'\beta'} = S^{u'\alpha'}_{u\alpha}S^{v\beta}_{v'\beta'}t^{u\alpha}_{v\beta} = \Delta^{u'\sigma}_{u}C^{\alpha'}_{\sigma\alpha}\Delta^{v\varepsilon}_{v'}C^{\beta}_{\varepsilon\beta'}t^{u\alpha}_{v\beta},$$

which implies

$$\begin{aligned}
\overset{*}{t}{}^{u'}_{v'} &= t^{i'}_{j'}\varepsilon^{\beta'}e_{\alpha'} = t^{u'\alpha'}_{v'\beta'}\varepsilon^{\beta'}e_{\alpha'} = S^{u'\alpha'}_{u\alpha}S^{v\beta}_{v'\beta'}t^{u\alpha}_{v\beta}\varepsilon^{\beta'}e_{\alpha'} \\
&= \Delta^{u'\sigma}_{u}C^{\alpha'}_{\sigma\alpha}\Delta^{v\varepsilon}_{v'}C^{\beta}_{\varepsilon\beta'}t^{u\alpha}_{v\beta}\varepsilon^{\beta'}e_{\alpha'} = \Delta^{u'\sigma}_{u}\Delta^{v\beta}_{v'}e_\sigma e_\alpha t^{u\alpha}_{v\beta} \\
&= \Delta^{u'\sigma}_{u}\Delta^{v\beta}_{v'}e_\sigma e_\alpha\overset{\varepsilon}{\mathfrak{I}}{}^{u}_{v}B^{\alpha}_{\varepsilon\beta} = \Delta^{u'\sigma}_{u}\Delta^{v\beta}_{v'}\overset{\varepsilon}{\mathfrak{I}}{}^{u}_{v}e_\sigma e_\beta e_\varepsilon \\
&= \overset{*}{S}{}^{u'}_{u}\overset{*}{S}{}^{v}_{v'}\overset{*}{t}{}^{u}_{v}
\end{aligned}$$

by virtue of (1.27).

Thus, we have

Theorem 1.8 *If Π is a regular integrable Π-structure on M_{mr}, then the pure tensor field t of type (r,s) on M_{mr} is a realization tensor of hypercomplex tensor $\overset{*}{t}$ in the $\mathfrak{A}$-holomorphic manifold $X_r(\mathfrak{A}_m)$.*

Remark 1.2 It is clear that the hypercomplex tensor field $\overset{*}{t}$ is not $\mathfrak{A}$-holomorphic, in general. In the next section, we will study a realization of the $\mathfrak{A}$-holomorphic tensor $\overset{*}{t}$.

1.6 Realizations of Holomorphic Tensors

Let $\mathfrak{A}_m$ be a Frobenius hypercomplex algebra and $\overset{*}{t} \in \mathfrak{I}^r_s(X_r(\mathfrak{A}_m))$ be a hypercomplex tensor field on $X_r(\mathfrak{A}_m)$. As previously mentioned, the realization of such a tensor field is a pure tensor field $t \in \mathfrak{I}^r_s(M_{mr})$, and $\overset{*}{t}$ is not holomorphic in general. To investigate a holomorphic algebraic tensor field $\overset{*}{t}$, we consider the Tachibana $\underset{\alpha}{\Phi_\varphi}$-operators on M_{mr}

associated with the regular Π-structure and applied to a pure tensor field t of type (r, s) [77, 86] (see also [52]):

$$\begin{aligned}
&\left(\Phi_{\underset{\alpha}{\varphi}} t\right)(X, Y_1, \ldots, Y_s, \xi^1, \ldots, \xi^r) \\
&= \left(\underset{\alpha}{\varphi} X\right) t(Y_1, \ldots, Y_s, \xi^1, \ldots, \xi^r) - X t\left(\underset{\alpha}{\varphi} Y_1, \ldots, Y_s, \xi^1, \ldots, \xi^r\right) \\
&\quad + \sum_{\lambda=1}^{s} t\left(Y_1, \ldots, \left(L_{Y_\lambda} \underset{\alpha}{\varphi}\right) X, \ldots, Y_s, \xi^1, \ldots, \xi^r\right) \\
&\quad - \sum_{\mu=1}^{r} t\left(Y_1, \ldots, Y_s, \xi^1, \ldots, L_{\varphi X} \xi^\mu - L_X\left(\xi^\mu \circ \underset{\alpha}{\varphi}\right), \ldots, \xi^r\right),
\end{aligned} \tag{1.34}$$

where $\Phi_{\underset{\alpha}{\varphi}} t \in \Im^r_{s+1}(M_{mr})$; $X, Y_\lambda \in \Im^1_0(M_{mr})$, $\xi^\mu \in \Im^0_1(M_{mr})$, $\lambda = 1, \ldots, s$; $\mu = 1, \ldots, r$; and L_Y is the Lie derivation with respect to Y.

In particular, if $t \in \Im^1_s(M_{mr})$, that is

$$\begin{aligned}
\varphi(t(Y_1, Y_2, \ldots, Y_s)) &= t(\varphi Y_1, Y_2, \ldots, Y_s) \\
&\;\;\vdots \\
&= t(Y_1, Y_2, \ldots, \varphi Y_s),
\end{aligned}$$

then from (1.34), we have

$$\left(\Phi_{\underset{\alpha}{\varphi}} t\right)(X, Y_1, Y_2, \ldots, Y_s) = -\left(L_{t(Y_1, Y_2, \ldots, Y_s)} \underset{\alpha}{\varphi}\right) X + \sum_{\lambda=1}^{s} t\left(Y_1, Y_2, \ldots, \left(L_{Y_\lambda} \underset{\alpha}{\varphi}\right) X, \ldots, Y_s\right).$$

Also, if $\omega \in \Im^0_s(M_{mr})$, then from (1.34), we have

$$\begin{aligned}
\left(\Phi_{\underset{\alpha}{\varphi}} \omega\right)(X, Y_1, \ldots, Y_s) &= \left(\underset{\alpha}{\varphi} X\right)(\omega(Y_1, \ldots, Y_s)) - X\left(\omega\left(\underset{\alpha}{\varphi} Y_1, \ldots, Y_s\right)\right) \\
&\quad + \sum_{\lambda=1}^{s} \omega\left(Y_1, \ldots, \left(L_{Y_\lambda} \underset{\alpha}{\varphi}\right) X, \ldots, Y_s\right) \\
&= \left(L_{\underset{\alpha}{\varphi} X} \omega - L_X\left(\omega \circ \underset{\alpha}{\varphi}\right)\right)(Y_1, \ldots, Y_s),
\end{aligned}$$

where $\omega \circ \varphi$ is defined by

$$\begin{aligned}
(\omega \circ \varphi)(Y_1, \ldots, Y_s) &= \omega(\varphi Y_1, Y_2, \ldots, Y_s) \\
&\;\;\vdots \\
&= \omega(Y_1, Y_2, \ldots, \varphi Y_s).
\end{aligned}$$

By setting $X = \partial_k, Y_\lambda = \partial_{j_\lambda}, \xi^\mu = dx^{i_\mu}$ in Eq. (1.34), we see that the components $\left(\underset{\alpha}{\Phi_\varphi} t\right)^{i_1 \ldots i_r}_{j_1 \ldots j_s}$ of $\underset{\alpha}{\Phi_\varphi} t$ with respect to local coordinate system $x^1, \ldots, x^n$ may be expressed as follows:

$$\begin{aligned}\left(\underset{\alpha}{\Phi_\varphi} t\right)^{i_1 \cdots i_r}_{k j_1 \cdots j_s} &= \underset{\alpha}{\varphi}{}^m_k \partial_m t^{i_1 \cdots i_r}_{j_1 \cdots j_s} - \partial_k \left(t \circ \underset{\alpha}{\varphi}\right)^{i_1 \cdots i_r}_{j_1 \cdots j_s} + \sum_{\lambda=1}^{s} \left(\partial_{j_\lambda} \underset{\alpha}{\varphi}{}^m_k\right) t^{i_1 \cdots i_r}_{j_1 \cdots m \cdots j_s} \\ &+ \sum_{\mu=1}^{r} \left(\partial_k \underset{\alpha}{\varphi}{}^{i_\mu}_m - \partial_m \underset{\alpha}{\varphi}{}^{i_\mu}_k\right) t^{i_1 \cdots m \cdots i_r}_{j_1 \cdots j_s}\end{aligned} \tag{1.35}$$

where

$$\begin{aligned}(t \circ \underset{\alpha}{\varphi})^{i_1 \cdots i_r}_{j_1 \cdots j_s} &= t^{i_1 \cdots i_r}_{m - j_s} \underset{\alpha}{\varphi}{}^m_{j_1} = \cdots = t^{i_1 \cdots i_r}_{j_1 \cdots m} \underset{\alpha}{\varphi}{}^m_{j_s} \\ &= t^{m \cdots i_r}_{j_1 \cdots j_s} \underset{\alpha}{\varphi}{}^{i_1}_m = \cdots = t^{i_s \cdots m}_{j_1 \cdots j_r} \underset{\alpha}{\varphi}{}^{i_r}_m .\end{aligned}$$

We note some important properties of operator Φ given by (1.34) to our further aims:

$$\begin{aligned}(i)(\underset{\alpha}{\Phi_\varphi} X)Y = -(L_X \underset{\alpha}{\varphi})Y, \quad (ii)(\underset{\alpha}{\Phi_\varphi} \underset{\beta}{\varphi})(X, Y) &= N_{\underset{\alpha}{\varphi}, \underset{\beta}{\varphi}}(X, Y) \\ &= [\underset{\alpha}{\varphi} X, \underset{\beta}{\varphi} Y] - \underset{\alpha}{\varphi}[X, \underset{\beta}{\varphi} Y] - \underset{\beta}{\varphi}[\underset{\alpha}{\varphi} X, Y] \\ &+ \underset{\alpha}{\varphi} \underset{\beta}{\varphi}[X, Y], X, Y \in \Im^1_0(M_{mr}),\end{aligned}$$

where $N_{\underset{\alpha}{\varphi}, \underset{\beta}{\varphi}}$ are the Nijenhuis tensors associated with Π [28].

The operator Φ given by (1.35) is first introduced by Tachibana [77]. The Tachibana operators and their generalizations were studied in [28, 52, 74, 86].

One can prove the following theorem (see [28, 60]):

Theorem 1.9 *Let Π be an integrable regular Π-structure on M_{mr}. The hypercomplex tensor field $\overset{*}{t} \in \Im^r_s(X_r(\mathfrak{A}_m))$ is $\mathfrak{A}$-holomorphic if and only if the pure tensor field $t \in \Im^r_s(M_{mr})$ (the realization of $\overset{*}{t}$) satisfies the equation.*

$$\underset{\alpha}{\Phi_\varphi} t = 0, \ \alpha = 1, \ldots, m,$$

where $\underset{\alpha}{\Phi_\varphi} t$ is the Tachibana operator defined by (1.35).

Proof Using adapted charts $\left(\partial_k \varphi^i_j = 0\right)$, by virtue of (1.18), from (1.35), we have ($i_a = u_a \alpha_a$, $j_b = v_b \beta_b$, $\mathrm{k} = w\gamma$; $a = 1, \ldots, r$; $b = 1, \ldots, s$).

$$\left(\Phi_{\underset{\alpha}{\varphi}} t\right)^{i_1\cdots i_r}_{j_1\cdots j_s} = \underset{\alpha}{\varphi}{}^m_k \partial_m t^{i_1\cdots i_r}_{j_1\cdots j_s} - \partial_k\left(t \circ \underset{\alpha}{\varphi}\right)^{i_1\cdots i_r}_{j_1\cdots j_s}$$
$$= \left(C^{\mu}_{\alpha\gamma}\partial_{w\mu}\overset{\lambda}{\mathfrak{I}}{}^{u_1\cdots u_r}_{v_1\cdots v_s} - C^{\lambda}_{\alpha\sigma}\partial_{w\gamma}\overset{\sigma}{\mathfrak{I}}{}^{u_1\cdots u_r}_{v_1\cdots v_s}\right)B^{\alpha_1\cdots\alpha_r}_{\lambda\beta_1\cdots\beta_s} = 0.$$

From B_3 (see Sect. 1.5), we see that the condition $\Phi_{\underset{\alpha}{\varphi}} t = 0$ is equivalent to the condition

$$C^{\mu}_{\alpha\gamma}\partial_{w\mu}\overset{\lambda}{\mathfrak{I}}{}^{u_1\cdots u_r}_{v_1\cdots v_s} = C^{\lambda}_{\alpha\sigma}\partial_{w\gamma}\overset{\sigma}{\mathfrak{I}}{}^{u_1\cdots u_r}_{v_1\cdots v_s}.$$

Using (1.13), we see that the last condition is the $\mathfrak{A}$-holomorphicity condition of $\overset{*}{t}{}^{u_1\cdots u_r}_{v_1\cdots v_s} = \overset{\sigma}{\mathfrak{I}}{}^{u_1\cdots u_r}_{v_1\cdots v_s} e_\sigma$ with respect to the local coordinates $z^u = x^{u\alpha}e_\alpha$ in $X_r(\mathfrak{A}_m)$. Thus, the proof is complete.

Remark 1.3 Let Π be a *nonintegrable* regular Π-structure on M_{mr}. Then if $t \in Ker\, \Phi_{\underset{\alpha}{\varphi}}$, we say that t is an *almost $\mathfrak{A}$-holomorphic$\mathfrak{A}-$* tensor field.

An *infinitesimal automorphism* of a regular Π-structure on M_{mr} is a vector field X such that $L_X \underset{\alpha}{\varphi} = 0$, $\alpha = 1, ..., m$, where L_X denotes the Lie differentiation with respect to $X \in \mathfrak{I}^1_0(M_{mr})$. From Theorem 1.9 and the property $i)$ of Tachibana operator, we have

Theorem 1.10 *Let on M_{mr} be given the integrable regular Π-structure. A vector field X is an infinitesimal automorphism if and only if X is $\mathfrak{A}$-holomorphic.*

Also we have

Theorem 1.11 *If ω be an exact 1-form, i.e. $\omega = df$, $f \in \mathfrak{I}^0_0(M_{mr})$, then ω is $\mathfrak{A}$-holomorphic if and only if the associated 1-forms $df \circ \underset{\alpha}{\varphi}$ are closed.*

Proof Let $\omega \in \mathfrak{I}^0_1(M)$. Using

$$(d\omega)(X, Y) = \frac{1}{2}\{X(\omega(Y)) - Y\omega(X) - \omega([X, Y])\}$$

for any $X, Y \in \mathfrak{I}^1_0 \in (M)$ and $\omega \in \mathfrak{I}^0_1(M)$, we have

$$(d\omega)\left(Y, \underset{\alpha}{\varphi} X\right) = \frac{1}{2}\left\{Y\left(\omega\left(\underset{\alpha}{\varphi} X\right)\right) - \left(\underset{\alpha}{\varphi} X\right)(\omega(Y)) - \omega\left(\left[Y, \underset{\alpha}{\varphi} X\right]\right)\right\}$$
$$= \frac{1}{2}\left\{Y\left(\omega\left(\underset{\alpha}{\varphi} X\right)\right) - \left(\underset{\alpha}{\varphi} X\right)(\omega(Y)) + \omega\left(\left[\underset{\alpha}{\varphi} X, Y\right]\right)\right\}$$

$$= \frac{1}{2}\Big\{ Y\Big(\omega\Big(\underset{\alpha}{\varphi} X\Big)\Big) - \Big(\underset{\alpha}{\varphi} X\Big)(\omega(Y)) + \omega\Big(\Big[\underset{\alpha}{\varphi} X, Y\Big] - \underset{\alpha}{\varphi}[X, Y]\Big) + \omega\Big(\underset{\alpha}{\varphi}[X, Y]\Big)\Big\} \tag{1.36}$$

From (1.34), we have

$$\Big(\underset{\alpha}{\Phi_{\varphi}}\omega\Big)(X, Y) = \Big(\underset{\alpha}{\varphi} X\Big)(\omega(Y)) - X\Big(\omega\Big(\underset{\alpha}{\varphi} Y\Big)\Big) - \omega\Big(\Big[\underset{\alpha}{\varphi} X, Y\Big] - \underset{\alpha}{\varphi}[X, Y]\Big) \tag{1.37}$$

Substituting (1.37) into (1.36), we have

$$\begin{aligned}
&(d\omega)\Big(Y, \underset{\alpha}{\varphi} X\Big) \\
&= \frac{1}{2}\Big\{-\Big(\underset{\alpha}{\Phi_{\varphi}}\omega\Big)(X, Y) + Y\Big(\omega\Big(\underset{\alpha}{\varphi} X\Big)\Big) - X\Big(\omega\Big(\underset{\alpha}{\varphi} Y\Big)\Big) + \omega\Big(\underset{\alpha}{\varphi}[X, Y]\Big)\Big\} \\
&= \frac{1}{2}\Big\{-\Big(\underset{\alpha}{\Phi_{\varphi}}\omega\Big)(X, Y) + Y\Big(\Big(\omega \circ \underset{\alpha}{\varphi}\Big)(X)\Big) - X\Big(\Big(\omega \circ \underset{\alpha}{\varphi}\Big)(Y)\Big) - \Big(\omega \circ \underset{\alpha}{\varphi}\Big)([Y, X])\Big\} \\
&= -\frac{1}{2}\Big(\underset{\alpha}{\Phi_{\varphi}}\omega\Big)(X, Y) + \Big(d\Big(\omega \circ \underset{\alpha}{\varphi}\Big)\Big)(Y, X).
\end{aligned}$$

From here, we see that equation $\underset{\alpha}{\Phi_{\varphi}}\omega = 0$ is equivalent to

$$\Big(d\Big(\omega \circ \underset{\alpha}{\varphi}\Big)\Big)(Y, X) = (d\omega)\Big(Y, \underset{\alpha}{\varphi} X\Big)$$

which for $\omega = df$ turns into the following simple relation:

$$\Big(d\Big(df \circ \underset{\alpha}{\varphi}\Big)\Big)(Y, X) = (d^2 f)\Big(Y, \underset{\alpha}{\varphi} X\Big) = 0,$$

i.e. $df \circ \underset{\alpha}{\varphi}$ is a closed 1-form. The proof is completed.

Let now $\omega \in \Im_1^0(M_{mr})$. Using (1.34), we have

$$\begin{aligned}
\Big(\underset{\alpha}{\Phi_{\varphi}}\omega\Big)(X, Y) &= \Big(\underset{\alpha}{\varphi} X\Big)(\omega(Y)) - X\Big(\omega(\underset{\alpha}{\varphi} Y)\Big) + \omega\Big(\Big(L_Y \underset{\alpha}{\varphi}\Big)X\Big) \\
&= \Big(L_{\underset{\alpha}{\varphi} X}\omega - L_X\Big(\omega \circ \underset{\alpha}{\varphi}\Big)\Big)Y
\end{aligned} \tag{1.38}$$

for any $X, Y \in \Im_0^1(M_{mr})$, where the associated 1-forms $\omega \circ \underset{\alpha}{\varphi}$ are defined by

$$\Big(\omega \circ \underset{\alpha}{\varphi}\Big)Y = \omega\Big(\underset{\alpha}{\varphi} Y\Big).$$

Theorem 1.12 *Let $\omega \in \Im_1^0(M_{2r})$ and φ be a complex or paracomplex structure on M_{2r}, i.e. $\varphi^2 = \mp id_{M_{2r}}$. Then the associated 1-form $\omega \circ \varphi$ is holomorphic if and only if $\omega \in$ Ker Φ_φ.*

Proof If we substitute $\omega \circ \varphi$ into ω and φX into X, then Eq. (1.38) may be written as

$$\begin{aligned}(\Phi_\varphi(\omega \circ \varphi))(\varphi X, Y) &= (L_{\varphi^2 X}(\omega \circ \varphi) - L_{\varphi X}(\omega \circ \varphi^2))Y \\ &= \mp(L_X(\omega \circ \varphi) - L_{\varphi X}\omega)Y \\ &= \pm(\Phi_\varphi \omega)(X, Y)\end{aligned}$$

or

$$((\Phi_\varphi(\omega \circ \varphi)) \circ \varphi)(X, Y) = \pm(\Phi_\varphi \omega)(X, Y),$$

from which by virtue of $\det \varphi \neq 0$, we see that $\Phi_\varphi(\omega \circ \varphi) = 0$ if and only if $\Phi_\varphi \omega = 0$. The proof is completed.

Theorem 1.13 *Let $\varphi^2 = \mp id_{M_{2r}}$. If $\omega \in \Im_1^0(M)$ is a holomorphic 1-form, then*

$$\omega \circ N_\varphi = 0,$$

where $\omega \circ N_\varphi$ is defined by $(\omega \circ N_\varphi)(X, Y) = \omega(N_\varphi(X, Y))$ and N_φ is the Nijenhuis tensor field associated with φ:

$$N_\varphi(X, Y) = [\varphi X, \varphi X] - \varphi[X, \varphi Y] - \varphi[\varphi X, Y] \mp [X, Y], \forall X, Y \in \Im_0^1(M_{2r}).$$

Proof immediately follows from Theorem 1.13 and the following formula:

$$\Phi_\varphi(\omega \circ \varphi) = (\Phi_\varphi \omega) \circ \varphi + \omega \circ N_\varphi.$$

1.7 Pure Connections

In this section, we always assume that the regular Π-structure is integrable, and we consider only local adapted coordinates with respect to the structure.

Let ∇ be a Π-connection on M_{mr}, i.e. $\nabla \underset{\sigma}{\varphi} = 0$ for any $\underset{\sigma}{\varphi} \in \Pi$. Since the components of $\underset{\sigma}{\varphi}$ with respect to the local adapted coordinates $x^1, ..., x^{mr}$ are constant, we have

$$\nabla \underset{\sigma}{\varphi} = 0 \Leftrightarrow \Gamma_{\mathrm{km}}^i \underset{\sigma}{\varphi}{}_j^m = \Gamma_{\mathrm{kj}}^m \underset{\sigma}{\varphi}{}_m^i . \tag{1.39}$$

By the same arguments as developed in Sect. 1.4, we see that the Π-connection has components of the form

$$\Gamma^{i}_{kj} = \Gamma^{u\alpha}_{w\gamma v\beta} = \overset{\sigma}{\tau}{}^{u}_{w\gamma v} C^{\alpha}_{\sigma\beta} \ (i = u\alpha,\ j = v\beta,\ \mathrm{k} = w\gamma), \tag{1.40}$$

where $\overset{\sigma}{\tau}{}^{u}_{w\gamma v} = \Gamma^{u\sigma}_{w\gamma v\beta}\varepsilon^{\beta}$ are any functions in the adapted chart $U \subset M_{mr}$. In fact, using contraction with ε^{β} and $m = t\varepsilon$, from (1.39), we obtain

$$\Gamma^{u\alpha}_{w\gamma t\varepsilon} \underset{\sigma}{\varphi}{}^{t\varepsilon}_{v\beta} = \Gamma^{t\varepsilon}_{w\gamma v\beta} \underset{\sigma}{\varphi}{}^{u\alpha}_{t\varepsilon},$$

$$\Gamma^{u\alpha}_{w\gamma t\varepsilon}\delta^{t}_{v}C^{\varepsilon}_{\sigma\beta} = \Gamma^{t\varepsilon}_{w\gamma v\beta}\delta^{u}_{t}C^{\alpha}_{\sigma\varepsilon},$$

$$\Gamma^{u\alpha}_{w\gamma v\sigma} = \Gamma^{u\varepsilon}_{w\gamma v\beta}\varepsilon^{\beta}C^{\alpha}_{\sigma\varepsilon} = \overset{\varepsilon}{\tau}{}^{u}_{w\gamma v}C^{\alpha}_{\varepsilon\sigma},$$

where $\overset{\varepsilon}{\tau}{}^{u}_{w\gamma v} = \Gamma^{u\varepsilon}_{w\gamma v\beta}\varepsilon^{\beta}$.

With Π-connection of type (2.40), we can associate a hypercomplex objects from $\mathfrak{A}_m$:

$$\begin{aligned} \overset{*}{\Gamma}{}^{u}_{wv} &= \Gamma^{u\alpha}_{w\gamma v\beta}\varepsilon^{\gamma}\varepsilon^{\beta}e_{\alpha} \\ &= \overset{\sigma}{\tau}{}^{u}_{w\gamma v}\varepsilon^{\gamma}e_{\sigma}. \end{aligned} \tag{1.41}$$

Definition 1.5 If the hypercomplex objects $\overset{*}{\Gamma}{}^{u}_{wv}$ satisfy the following condition:

$$\overset{*}{\Gamma}{}^{u'}_{w'v'} = \frac{\partial z^{u'}}{\partial z^{u}}\frac{\partial z^{w}}{\partial z^{w'}}\frac{\partial z^{v}}{\partial z^{v'}}\overset{*}{\Gamma}{}^{u}_{wv} + \frac{\partial^2 z^{u}}{\partial z^{v'}\partial z^{w'}}\frac{\partial z^{u'}}{\partial z^{u}},$$

i.e. if $\overset{*}{\Gamma}{}^{u}_{wv}$ are the components of the hypercomplex connection $\overset{*}{\nabla}$ in $X_r(\mathfrak{A}_m)$, then we say that the Π-connection ∇ is a pure connection.

Theorem 1.14 *Let Π be a regular integrable Π-structure on M_{mr}. The Π-connection ∇ is pure if and only if $\overset{\sigma}{\tau}{}^{u}_{w\gamma v}$ satisfies the condition*

$$\overset{\alpha}{\tau}{}^{u}_{w\gamma v} = \overset{\sigma}{\tau}{}^{u}_{wv}C^{\alpha}_{\sigma\gamma}, \tag{1.42}$$

where $\overset{\sigma}{\tau}{}^{u}_{wv} = \overset{\sigma}{\tau}{}^{u}_{w\eta v}\varepsilon^{\eta}$.

Proof Let $\Gamma^{i}_{kj} = \Gamma^{u\alpha}_{w\gamma v\beta} = \overset{\sigma}{\tau}{}^{u}_{w\gamma v}C^{\alpha}_{\sigma\beta}$ be the components of the Π-connection ∇. Then, taking account of the admissible transformation $\{\partial_i\} \to \{\partial_{i'}\}$ of adapted frames with matrix $\left(S^{i'}_{i}\right) = \left(\frac{\partial x^{i'}}{\partial x^{i}}\right) = \left(\frac{\partial x^{u'\alpha'}}{\partial x^{u\alpha}}\right)$, we have

$$\Gamma^{u'\alpha'}_{w'\gamma'\,v'\beta'} = \frac{\partial x^{u'\alpha'}}{\partial x^{u\alpha}}\frac{\partial x^{w\gamma}}{\partial x^{w'\gamma'}}\frac{\partial x^{v\beta}}{\partial x^{v'\beta'}}\Gamma^{u\alpha}_{w\gamma v\beta} + \frac{\partial^2 x^{u\alpha}}{\partial x^{w'\gamma'}\partial x^{v'\beta'}}\frac{\partial x^{u'\alpha'}}{\partial x^{u\alpha}}.$$

After contraction with $\varepsilon^{\gamma'}\varepsilon^{\beta'}e_{\alpha'}$, by virtue of (1.21), (1.22) and (1.40), we obtain

$$\begin{aligned}\overset{*}{\Gamma}{}^{u'}_{w'v'} &= \Gamma^{u'\alpha'}_{w'\gamma' v'\beta'}\varepsilon^{\gamma'}\varepsilon^{\beta'}e_{\alpha'}\\ &= \overset{\sigma}{\Delta}{}^{u'}_{u}C^{\alpha'}_{\sigma\alpha}\overset{\varepsilon}{\Delta}{}^{w}_{w'}C^{\gamma}_{\varepsilon\gamma'}\overset{\theta}{\Delta}{}^{v}_{v'}C^{\beta}_{\theta\beta'}\Gamma^{u\alpha}_{w\gamma v\beta}\varepsilon^{\gamma'}\varepsilon^{\beta'}e_{\alpha'}\\ &+ \overset{\sigma}{\Delta}{}^{u'}_{u}C^{\alpha'}_{\sigma\alpha}\left(\frac{\partial}{\partial x^{w'\gamma'}}\left(\overset{\varepsilon}{\Delta}{}^{u}_{v'}C^{\alpha}_{\varepsilon\beta'}\right)\right)\varepsilon^{\gamma'}\varepsilon^{\beta'}e_{\alpha'}\end{aligned}$$

or

$$\begin{aligned}\overset{*}{\Gamma}{}^{u'}_{w'v'} &= \overset{\sigma}{\Delta}{}^{u'}_{u}\overset{\varepsilon}{\Delta}{}^{w}_{w'}\overset{\theta}{\Delta}{}^{v}_{v'}e_{\sigma}e_{\alpha}\delta^{\gamma}_{\varepsilon}\delta^{\beta}_{\theta}\overset{\omega}{\tau}{}^{u}_{w\gamma v}C^{\alpha}_{\omega\beta} + \overset{\sigma}{\Delta}{}^{u'}_{u}e_{\sigma}\varepsilon^{\gamma'}\left(\frac{\partial}{\partial x^{w'\gamma'}}\left(\overset{\alpha}{\Delta}{}^{u}_{v'}\right)\right)e_{\alpha}\\ &= \overset{*}{S}{}^{u'}_{u}\overset{*}{S}{}^{v}_{v'}\overset{\gamma}{\Delta}{}^{w}_{w'}\overset{\theta}{\tau}{}^{u}_{w\gamma v}e_{\theta} + \overset{*}{S}{}^{u'}_{u}\varepsilon^{\gamma'}\left(\frac{\partial}{\partial x^{w'\gamma'}}\left(\overset{\alpha}{\Delta}{}^{u}_{v'}\right)\right)e_{\alpha},\end{aligned}$$

where

$$\overset{*}{S}{}^{u'}_{u} = \frac{\partial z^{u'}}{\partial z^{u}},\ S^{u}_{u'} = \frac{\partial z^{u}}{\partial z^{u'}},\ z^{u} = x^{u\alpha}e_{\alpha}.$$

Using (1.14), we see that

$$\varepsilon^{\gamma'}\left(\frac{\partial}{\partial x^{w'\gamma'}}\left(\overset{\alpha}{\Delta}{}^{u}_{v'}\right)\right)e_{\alpha} = \frac{\partial^2 z^{u}}{\partial z^{w'}\partial z^{v'}}.$$

Thus, we have

$$\overset{*}{\Gamma}{}^{u'}_{w'v'} = \frac{\partial z^{u'}}{\partial z^{u}}\overset{\gamma}{\Delta}{}^{w}_{w'}\frac{\partial z^{v}}{\partial z^{v'}}\overset{\theta}{\tau}{}^{u}_{w\gamma v}e_{\theta} + \frac{\partial^2 z^{u}}{\partial z^{v'}\partial z^{w'}}\frac{\partial z^{u'}}{\partial z^{u}}. \tag{1.43}$$

From (1.43), we see that $\overset{*}{\nabla}$ with components $\overset{*}{\Gamma}{}^{u}_{wv}$ is hypercomplex connection on $X_r(\mathfrak{A}_m)$ if and only if $\overset{\theta}{\tau}{}^{u}_{w\gamma v}$ satisfies the condition $\overset{\alpha}{\tau}{}^{u}_{w\gamma v} = \overset{\sigma}{\tau}{}^{u}_{wv}C^{\alpha}_{\sigma\gamma}$. In fact, substituting $\overset{\alpha}{\tau}{}^{u}_{w\gamma v} = \overset{\sigma}{\tau}{}^{u}_{wv}C^{\alpha}_{\sigma\gamma}$ into (1.41) and (1.43), we have

$$\overset{*}{\Gamma}{}^{u}_{wv} = \Gamma^{u\alpha}_{w\gamma v\beta}\varepsilon^{\gamma}\varepsilon^{\beta}e_{\alpha} = \overset{\sigma}{\tau}{}^{u}_{w\gamma v}\varepsilon^{\gamma}e_{\sigma} = \overset{\theta}{\tau}{}^{u}_{wv}C^{\sigma}_{\theta\gamma}\varepsilon^{\gamma}e_{\sigma} = \overset{\theta}{\tau}{}^{u}_{wv}\delta^{\sigma}_{\theta}e_{\sigma} = \overset{\theta}{\tau}{}^{u}_{wv}e_{\theta}$$

and

$$\begin{aligned}\overset{*}{\Gamma}{}^{u'}_{w'v'} &= \frac{\partial z^{u'}}{\partial z^{u}}\overset{\gamma}{\Delta}{}^{w}_{w'}\frac{\partial z^{v}}{\partial z^{v'}}\overset{\theta}{\tau}{}^{u}_{w\gamma v}e_{\theta} + \frac{\partial^2 z^{u}}{\partial z^{v'}\partial z^{w'}}\frac{\partial z^{u'}}{\partial z^{u}}\\ &= \frac{\partial z^{u'}}{\partial z^{u}}\overset{\gamma}{\Delta}{}^{w}_{w'}\frac{\partial z^{v}}{\partial z^{v'}}\overset{\sigma}{\tau}{}^{u}_{wv}C^{\theta}_{\sigma\gamma}e_{\theta} + \frac{\partial^2 z^{u}}{\partial z^{v'}\partial z^{w'}}\frac{\partial z^{u'}}{\partial z^{u}}\\ &= \frac{\partial z^{u'}}{\partial z^{u}}\frac{\partial z^{w}}{\partial z^{w'}}\frac{\partial z^{v}}{\partial z^{v'}}\overset{\sigma}{\tau}{}^{u}_{wv}e_{\sigma} + \frac{\partial^2 z^{u}}{\partial z^{v'}\partial z^{w'}}\frac{\partial z^{u'}}{\partial z^{u}}\end{aligned}$$

$$= \frac{\partial z^{u'}}{\partial z^u}\frac{\partial z^w}{\partial z^{w'}}\frac{\partial z^v}{\partial z^{v'}}\overset{*}{\Gamma}{}^u_{wv} + \frac{\partial^2 z^u}{\partial z^{v'}\partial z^{w'}}\frac{\partial z^{u'}}{\partial z^u}.$$

Conversely, comparing (1.43) with the last equation, we get

$$\overset{\gamma}{\Delta}{}^w_{w'}\overset{\theta}{\tau}{}^u_{w\gamma v}e_\theta = \frac{\partial z^w}{\partial z^{w'}}\overset{*}{\Gamma}{}^u_{wv} \Leftrightarrow \overset{\gamma}{\Delta}{}^w_{w'}\overset{\theta}{\tau}{}^u_{w\gamma v}e_\theta e_\gamma = \frac{\partial z^w}{\partial z^{w'}}\overset{*}{\Gamma}{}^u_{wv}e_\gamma,$$

$$\frac{\partial z^w}{\partial z^{w'}}\overset{\theta}{\tau}{}^u_{w\gamma v}e_\theta = \frac{\partial z^w}{\partial z^{w'}}\overset{\sigma}{\tau}{}^u_{w\eta v}\varepsilon^\eta e_\sigma e_\gamma \Leftrightarrow \overset{\theta}{\tau}{}^u_{w\gamma v}e_\theta = \overset{\sigma}{\tau}{}^u_{w\eta v}\varepsilon^\eta e_\sigma e_\gamma,$$

$$\overset{\theta}{\tau}{}^u_{w\gamma v}e_\theta = \overset{\sigma}{\tau}{}^u_{w\eta v}\varepsilon^\eta C^\theta_{\sigma\gamma}e_\theta \Leftrightarrow \overset{\theta}{\tau}{}^u_{w\gamma v} = \overset{\sigma}{\tau}{}^u_{w\eta v}\varepsilon^\eta C^\theta_{\sigma\gamma}.$$

Thus, $\overset{\theta}{\tau}{}^u_{w\gamma v} = \overset{\sigma}{\tau}{}^u_{wv}C^\theta_{\sigma\gamma}$, where $\overset{\sigma}{\tau}{}^u_{wv} = \overset{\sigma}{\tau}{}^u_{w\eta v}\varepsilon^\eta$. The proof is completed.
Using (1.42), we get from (1.40) and (1.41), respectively

$$\Gamma^i_{kj} = \Gamma^{u\alpha}_{w\gamma v\beta} = \overset{\sigma}{\tau}{}^u_{wv}C^\mu_{\sigma\gamma}C^\alpha_{\mu\beta} = \overset{\sigma}{\tau}{}^u_{wv}B^\alpha_{\sigma\gamma\beta} \tag{1.44}$$

and

$$\overset{*}{\Gamma}{}^u_{wv} = \overset{\sigma}{\tau}{}^u_{wv}e_\sigma, \tag{1.45}$$

where $B^\alpha_{\sigma\gamma\beta}$ is the Kruchkovich tensor.
Thus, we have

Theorem 1.15 *A pure* Π*-connection* ∇ *has the components* (1.44) *with respect to the adapted coordinates.*

Theorem 1.16 *A pure* Π*-connection* ∇ *is a realization of the hypercomplex connection* $\overset{*}{\nabla}$ *with components* (1.45).

From Theorem 1.15 and (1.32), it follows that the pure Π-connection as a pure tensor fields of type (1,2) is defined by

$$\Gamma^i_{km}\underset{\alpha}{\varphi}{}^m_j = \Gamma^m_{kj}\underset{\alpha}{\varphi}{}^i_m = \Gamma^i_{mj}\underset{\alpha}{\varphi}{}^m_k, \alpha = 1, ..., m$$

with respect to the *adapted* charts. But the pure tensor fields of type (1,2) (see (1.23)) are defined by a similiar equation with respect to the *arbitrary* charts.

1.8 Torsion Tensors of Pure Connections

Let S be a torsion tensor of pure Π-connection ∇. Since $B^\alpha_{\sigma\gamma\beta} = B^\alpha_{\sigma\beta\gamma}$ (see Sect. 1.5), from (1.44), we have

$$S^i_{kj} = \Gamma^i_{kj} - \Gamma^i_{jk} = (\overset{\sigma}{\tau}{}^{u}_{wv} - \overset{\sigma}{\tau}{}^{u}_{vw}) B^{\alpha}_{\sigma\gamma\beta}, \tag{1.46}$$

i.e. S is a pure tensor (see (1.32)).

Conversely, we now assume that S is a pure torsion tensor of the Π-connection. Then, by virtue of (1.32), we have

$$S^i_{kj} = S^{u\alpha}_{w\gamma v\beta} = \overset{\lambda}{\sigma}{}^{u}_{wv} B^{\alpha}_{\lambda\gamma\beta}. \tag{1.47}$$

On the other hand, from (1.40), we have

$$S^i_{kj} = \Gamma^i_{kj} - \Gamma^i_{jk} = \overset{\lambda}{\tau}{}^{u}_{w\gamma v} C^{\alpha}_{\lambda\beta} - \overset{\lambda}{\tau}{}^{u}_{v\beta w} C^{\alpha}_{\lambda\gamma}. \tag{1.48}$$

From (1.47) and (1.48), by virtue of contraction with ε^{γ}, we have

$$\overset{\alpha}{\tau}{}^{u}_{v\beta w} = \left(\overset{\lambda}{\tau}{}^{u}_{w\gamma v}\varepsilon^{\gamma} - \overset{\lambda}{\sigma}{}^{u}_{wv}\right) C^{\alpha}_{\lambda\beta},$$

which shows the condition of type (1.42) is true, i.e. the Π-connection is pure.

Thus, we have

Theorem 1.17 *The Π-connection ∇ is pure if and only if its torsion tensor is pure.*

For the pure Π-connection, by virtue of (1.33), (1.45) and (1.46), we obtain

$$\overset{*}{S}{}^{u}_{wv} = (\overset{\sigma}{\tau}{}^{u}_{wv} - \overset{\sigma}{\tau}{}^{u}_{vw}) e_{\sigma} = \overset{*}{\Gamma}{}^{u}_{wv} - \overset{*}{\Gamma}{}^{u}_{vw}.$$

Thus, we have.

Theorem 1.18. *The pure torsion tensor field S of the Π-connection ∇ is realization of the hypercomplex torsion tensor $\overset{*}{S}$ of hypercomplex connection $\overset{*}{\nabla}$.*

It is well known that the zero tensor field is pure. Therefore, we have

Theorem 1.19 *A torsion-free Π-connection ∇ is pure.*

Also, from Theorems 1.18 and 1.19, we have

Theorem 1.20 *If* ∇ *is a torsion-free* Π*-connection, then* $\overset{*}{\nabla}$ *with components* $\overset{*}{\Gamma}{}^{u}_{wv} = \overset{\sigma}{\tau}{}^{u}_{wv} e_{\sigma}$ *is also a torsion-free connection.*

1.9 Holomorphic Connections

Let R be a curvature tensor of the pure Π-connection ∇. Using (1.44) and the properties of Kruchkovich tensors, we have

$$\begin{aligned} R^{i}_{jkl} &= R^{u\alpha}_{v\beta w\gamma t\delta} \\ &= \partial_{v\beta}\Gamma^{u\alpha}_{w\gamma t\delta} - \partial_{w\gamma}\Gamma^{u\alpha}_{v\beta t\delta} + \Gamma^{u\alpha}_{v\beta x\varepsilon}\Gamma^{x\varepsilon}_{w\gamma t\delta} - \Gamma^{u\alpha}_{w\gamma x\varepsilon}\Gamma^{x\varepsilon}_{v\beta t\delta} \\ &= \partial_{v\beta}\Big(\overset{\sigma}{\tau}{}^{u}_{wt} B^{\alpha}_{\sigma\gamma\delta}\Big) - \partial_{w\gamma}\Big(\overset{\sigma}{\tau}{}^{u}_{vt} B^{\alpha}_{\sigma\beta\delta}\Big) + \overset{\sigma}{\tau}{}^{u}_{vx} B^{\alpha}_{\sigma\beta\varepsilon}\overset{\theta}{\tau}{}^{x}_{wt} B^{\varepsilon}_{\theta\gamma\delta} - \overset{\sigma}{\tau}{}^{u}_{wx} B^{\alpha}_{\sigma\gamma\varepsilon}\overset{\theta}{\tau}{}^{x}_{vt} B^{\varepsilon}_{\theta\beta\delta} \\ &= \Big(\partial_{v\beta}\overset{\sigma}{\tau}{}^{u}_{wt}\Big) B^{\alpha}_{\sigma\gamma\delta} - \Big(\partial_{w\gamma}\overset{\sigma}{\tau}{}^{u}_{vt}\Big) B^{\alpha}_{\sigma\beta\delta} + \Big(\overset{\sigma}{\tau}{}^{u}_{vx}\overset{\theta}{\tau}{}^{x}_{wt} - \overset{\sigma}{\tau}{}^{u}_{wx}\overset{\theta}{\tau}{}^{x}_{vt}\Big) B^{\alpha}_{\sigma\gamma\theta\beta\delta}. \end{aligned} \tag{1.49}$$

We now assume that R is a pure tensor. Then by virtue of (1.32), we have

$$R^{i}_{jkl} = R^{u\alpha}_{v\beta w\gamma t\delta} = \overset{\lambda}{\rho}{}^{u}_{vwt} B^{\alpha}_{\lambda\beta\gamma\delta} \tag{1.50}$$

From (2.49) and (2.50), we obtain

$$\overset{\lambda}{\rho}{}^{u}_{vwt} B^{\alpha}_{\lambda\beta\gamma\delta} = \Big(\partial_{v\beta}\overset{\sigma}{\tau}{}^{u}_{wt}\Big) B^{\alpha}_{\sigma\gamma\delta} - \Big(\partial_{w\gamma}\overset{\sigma}{\tau}{}^{u}_{vt}\Big) B^{\alpha}_{\sigma\beta\delta} + \Big(\overset{\sigma}{\tau}{}^{u}_{vx}\overset{\theta}{\tau}{}^{x}_{wt} - \overset{\sigma}{\tau}{}^{u}_{wx}\overset{\theta}{\tau}{}^{x}_{vt}\Big) B^{\alpha}_{\sigma\gamma\theta\beta\delta}. \tag{1.51}$$

Using contraction with $\varepsilon^{\beta}\varepsilon^{\delta}$ and the properties of Kruchkovich tensors, from (2.51), we have

$$\begin{aligned} \overset{\lambda}{\rho}{}^{u}_{vwt} C^{\alpha}_{\lambda\gamma} &= \varepsilon^{\beta}\Big(\partial_{v\beta}\overset{\sigma}{\tau}{}^{u}_{wt}\Big) C^{\alpha}_{\sigma\gamma} - \partial_{w\gamma}\overset{\alpha}{\tau}{}^{u}_{vt} + \Big(\overset{\sigma}{\tau}{}^{u}_{vx}\overset{\theta}{\tau}{}^{x}_{wt} - \overset{\sigma}{\tau}{}^{u}_{wx}\overset{\theta}{\tau}{}^{x}_{vt}\Big) C^{\lambda}_{\sigma\theta} C^{\alpha}_{\lambda\gamma} \\ &\varepsilon^{\beta}\Big(\partial_{v\beta}\overset{\sigma}{\tau}{}^{u}_{wt}\Big) C^{\alpha}_{\sigma\gamma} - \partial_{w\gamma}\overset{\alpha}{\tau}{}^{u}_{vt} + \Big(\overset{\sigma}{\tau}{}^{u}_{vx}\overset{\theta}{\tau}{}^{x}_{wt} - \overset{\sigma}{\tau}{}^{u}_{wx}\overset{\theta}{\tau}{}^{x}_{vt}\Big) C^{\lambda}_{\sigma\theta} C^{\alpha}_{\lambda\gamma} \\ &= -\partial_{w\gamma}\overset{\alpha}{\tau}{}^{u}_{vt} + \Big(\varepsilon^{\beta}\partial_{v\beta}\overset{\lambda}{\tau}{}^{u}_{wt} + \overset{\sigma}{\tau}{}^{u}_{vx}\overset{\theta}{\tau}{}^{x}_{wt} C^{\lambda}_{\sigma\theta} - \overset{\sigma}{\tau}{}^{u}_{wx}\overset{\theta}{\tau}{}^{x}_{vt} C^{\lambda}_{\sigma\theta}\Big) C^{\alpha}_{\lambda\gamma}. \end{aligned}$$

From here, it follows that

$$\partial_{w\gamma}\overset{\alpha}{\tau}{}^{u}_{vt} = \overset{\lambda}{P}{}^{u}_{vwt} C^{\alpha}_{\lambda\gamma}, \tag{1.52}$$

where

$$\overset{\lambda}{P}{}^{u}_{vwt} = \varepsilon^{\beta}\partial_{v\beta}\overset{\lambda}{\tau}{}^{u}_{wt} + \overset{\sigma}{\tau}{}^{u}_{vx}\overset{\theta}{\tau}{}^{x}_{wt}C^{\lambda}_{\sigma\theta} - \overset{\sigma}{\tau}{}^{u}_{wx}\overset{\theta}{\tau}{}^{x}_{vt}C^{\lambda}_{\sigma\theta} - \overset{\lambda}{\rho}{}^{u}_{vwt}.$$

By virtue of (1.12), for fixed u, v, w and t, the condition (1.52) is the $\mathfrak{A}$-holomorphity condition of $\overset{*}{\Gamma}{}^{u}_{vt} = \overset{\alpha}{\tau}{}^{u}_{vt}e_{\alpha}$ with respect to the local coordinates $z^{u} = x^{u\alpha}e_{\alpha}$ in $X_r(\mathfrak{A}_m)$.

Conversely, if $\overset{*}{\Gamma}{}^{u}_{vt} = \overset{\alpha}{\tau}{}^{u}_{vt}e_{\alpha}$ is a $\mathfrak{A}$-holomorphic connection, then from Theorems 1.4 and 1.1 ($\widetilde{C}_{\alpha} = C_{\alpha}$) and (1.12), we obtain the condition of type (1.52). Using (1.52), from (1.49), we have

$$\begin{aligned} R^{i}_{jkl} &= R^{u\alpha}_{v\beta w\gamma t\delta} \\ &= \left(\overset{\lambda}{P}{}^{u}_{wvt}C^{\sigma}_{\lambda\beta}\right)B^{\alpha}_{\sigma\gamma\delta} - \left(\overset{\lambda}{P}{}^{u}_{vwt}C^{\sigma}_{\lambda\gamma}\right)B^{\alpha}_{\sigma\beta\delta} + \left(\overset{\sigma}{\tau}{}^{u}_{vx}\overset{\theta}{\tau}{}^{x}_{wt} - \overset{\sigma}{\tau}{}^{u}_{wx}\overset{\theta}{\tau}{}^{x}_{vt}\right)B^{\alpha}_{\sigma\gamma\theta\beta\delta} \\ &= \overset{\lambda}{P}{}^{u}_{wvt}B^{\alpha}_{\lambda\beta\gamma\delta} - \overset{\lambda}{P}{}^{u}_{vwt}B^{\alpha}_{\lambda\gamma\beta\delta} + \left(\overset{\sigma}{\tau}{}^{u}_{vx}\overset{\theta}{\tau}{}^{x}_{wt} - \overset{\sigma}{\tau}{}^{u}_{wx}\overset{\theta}{\tau}{}^{x}_{vt}\right)C^{\lambda}_{\sigma\theta}B^{\alpha}_{\lambda\gamma\beta\delta} \\ &= \overset{\lambda}{\rho}{}^{u}_{vwt}B^{\alpha}_{\lambda\beta\gamma\delta}, \end{aligned} \tag{1.53}$$

where

$$\overset{\lambda}{\rho}{}^{u}_{vwt} = \overset{\lambda}{P}{}^{u}_{wvt} - \overset{\lambda}{P}{}^{u}_{vwt} + \left(\overset{\sigma}{\tau}{}^{u}_{vx}\overset{\theta}{\tau}{}^{x}_{wt} - \overset{\sigma}{\tau}{}^{u}_{wx}\overset{\theta}{\tau}{}^{x}_{vt}\right)C^{\lambda}_{\sigma\theta}.$$

From (1.32) and (1.53), we see that R is a pure curvature tensor.

Thus, we have

Theorem 1.21 [28, 82] *Let $\overset{*}{\nabla}$ be a hypercomplex connection on $X_r(\mathfrak{A}_m)$ and ∇ its realization on M_{mr}. The curvature tensor R of ∇ is pure if and only if $\overset{*}{\nabla}$ is an $\mathfrak{A}$-holomorphic connection.*

1.10 Holomorphic Curvature Tensor

Let now R be a pure tensor of pure Π-connection ∇. Using (1.3) and (1.49), we have

$$\begin{aligned} \overset{*}{R}{}^{u}_{vwt} &= R^{u\alpha}_{v\beta w\gamma t\delta}\varepsilon^{\beta}\varepsilon^{\gamma}\varepsilon^{\delta}e_{\alpha} \\ &= \varepsilon^{\beta}\left(\partial_{v\beta}\overset{\alpha}{\tau}{}^{u}_{wt}\right)e_{\alpha} - \varepsilon^{\gamma}\left(\partial_{w\gamma}\overset{\alpha}{\tau}{}^{u}_{vt}\right)e_{\alpha} + \left(\overset{\sigma}{\tau}{}^{u}_{vx}\overset{\theta}{\tau}{}^{x}_{wt} - \overset{\sigma}{\tau}{}^{u}_{wx}\overset{\theta}{\tau}{}^{x}_{vt}\right)C^{\alpha}_{\sigma\theta}e_{\alpha}. \end{aligned}$$

Since, $C^{\alpha}_{\sigma\theta}e_{\alpha} = e_{\sigma}e_{\theta}$, by virtue of (1.14) and (1.45), we have

$$\overset{*}{R}{}^{u}_{vwt} = \partial_{v}\overset{*}{\Gamma}{}^{u}_{wt} - \partial_{w}\overset{*}{\Gamma}{}^{u}_{vt} + \overset{*}{\Gamma}{}^{u}_{vx}\overset{*}{\Gamma}{}^{x}_{wt} - \overset{*}{\Gamma}{}^{u}_{wx}\overset{*}{\Gamma}{}^{x}_{vt},$$

i.e. $\overset{*}{R}$ is a curvature tensor of $\overset{*}{\Gamma}$.

Thus, we have

Theorem 1.22 *Let ∇ be a realization of the $\mathfrak{A}$-holomorphic connection $\overset{*}{\nabla}$. The hypercomplex components $\overset{*}{R}{}_{vwt}^{\ \ u}$ associated with the pure curvature tensor R are components of curvature tensor field $\overset{*}{R}$ of $\overset{*}{\nabla}$.*

Let now $\overset{*}{\nabla}$ be an $\mathfrak{A}$-holomorphic hypercomplex connection on $X_r(\mathfrak{A}_m)$ and a pure Π-connection ∇ its realization on M_{mr}. From Theorem 1.21 we see that the curvature tensor R of ∇ is pure. Since the curvature tensor R is pure, we can apply the Tachibana operator $\Phi_{\underset{\alpha}{\varphi}}$ to R:

$$(\Phi_{\underset{\alpha}{\varphi}} R)(X, Y_1, Y_2, Y_3) = -\left(L_{R(Y_1,Y_2,Y_3)} \underset{\alpha}{\varphi}\right)X + R\left(\left(L_{Y_1} \underset{\alpha}{\varphi}\right)X, Y_2, Y_3\right)$$
$$+ R\left(Y_1, \left(L_{Y_2} \underset{\alpha}{\varphi}\right)X, Y_3\right) + R\left(Y_1, Y_2, \left(L_{Y_3} \underset{\alpha}{\varphi}\right)X\right),$$

where $\underset{\alpha}{\varphi} R(Y_1, Y_2, Y_3) = R\left(\underset{\alpha}{\varphi} Y_1, Y_2, Y_3\right) = R\left(Y_1, \underset{\alpha}{\varphi} Y_2, Y_3\right) = R\left(Y_1, Y_2, \underset{\alpha}{\varphi} Y_3\right)$.

Now we proved that if a torsion tensor of Π-connection ∇ is pure, then

$$-(L_Y \underset{\alpha}{\varphi})X = \nabla_{\underset{\alpha}{\varphi} X} Y - \underset{\alpha}{\varphi}(\nabla_X Y).$$

In fact, from $\nabla \underset{\alpha}{\varphi} = 0$ and $\underset{\alpha}{\varphi} S(X, Y) = S(\underset{\alpha}{\varphi} X, Y) = S(X, \underset{\alpha}{\varphi} Y)$, where $S(X, Y) = \nabla_X Y - \nabla_Y X - [X, Y]$, we obtain

$$\begin{aligned} -\left(L_Y \underset{\alpha}{\varphi}\right)X &= \left[\underset{\alpha}{\varphi} X, Y\right] - \underset{\alpha}{\varphi}[X, Y] \\ &= \nabla_{\underset{\alpha}{\varphi} X} Y - \nabla_Y \underset{\alpha}{\varphi} X - S\left(\underset{\alpha}{\varphi} X, Y\right) - \underset{\alpha}{\varphi}(\nabla_X Y - \nabla_Y X - S(X, Y)) \\ &= \nabla_{\underset{\alpha}{\varphi} X} Y - \underset{\alpha}{\varphi}(\nabla_X Y) - \left(\nabla \underset{\alpha}{\varphi}\right)(Y, X) + \underset{\alpha}{\varphi} S(X, Y) - S\left(\underset{\alpha}{\varphi} X, Y\right) \\ &= \nabla_{\underset{\alpha}{\varphi} X} Y - \underset{\alpha}{\varphi}(\nabla_X Y), \end{aligned}$$

Now using the purity conditions of R and S, and also the last formula, we have

$$\left(\Phi_{\underset{\alpha}{\varphi}} R\right)(X, Y_1, Y_2, Y_3) = -\left(L_{R(Y_1,Y_2,Y_3)} \underset{\alpha}{\varphi}\right)X + R\left(\left(L_{Y_1} \underset{\alpha}{\varphi}\right)X, Y_2, Y_3\right)$$
$$+ R\left(Y_1, \left(L_{Y_2} \underset{\alpha}{\varphi}\right)X, Y_3\right) + R\left(Y_1, Y_2, \left(L_{Y_3} \underset{\alpha}{\varphi}\right)X\right)$$

$$= \left[\underset{\alpha}{\varphi} X, R(Y_1, Y_2, Y_3)\right] - \underset{\alpha}{\varphi}[X, R(Y_1, Y_2, Y_3)]$$
$$+ R\left(\left(L_{Y_1} \underset{\alpha}{\varphi}\right)X, Y_2, Y_3\right) + R\left(Y_1, \left(L_{Y_2} \underset{\alpha}{\varphi}\right)X, Y_3\right) + R\left(Y_1, Y_2, \left(L_{Y_3} \underset{\alpha}{\varphi}\right)X\right)$$
$$= \nabla_{\underset{\alpha}{\varphi} X} R(Y_1, Y_2, Y_3) - \nabla_{R(Y_1,Y_2,Y_3)} \underset{\alpha}{\varphi} X - S\left(\underset{\alpha}{\varphi} X, R(Y_1, Y_2, Y_3)\right)$$
$$- \underset{\alpha}{\varphi}(\nabla_X R(Y_1, Y_2, Y_3) - \nabla_{R(Y_1,Y_2,Y_3)} X - S(X, R(Y_1, Y_2, Y_3)))$$
$$+ R\left(\left(L_{Y_1} \underset{\alpha}{\varphi}\right)X, Y_2, Y_3\right) + R\left(Y_1, \left(L_{Y_2} \underset{\alpha}{\varphi}\right)X, Y_3\right) + R\left(Y_1, Y_2, \left(L_{Y_3} \underset{\alpha}{\varphi}\right)X\right)$$
$$= \left(\nabla_{\underset{\alpha}{\varphi} X} R\right)(Y_1, Y_2, Y_3) + R\left(\nabla_{\underset{\alpha}{\varphi} X} Y_1, Y_2, Y_3\right) + R\left(Y_1, \nabla_{\underset{\alpha}{\varphi} X} Y_2, Y_3\right)$$
$$+ R\left(Y_1, Y_2, \nabla_{\underset{\alpha}{\varphi} X} Y_3\right) - \left(\nabla_{R(Y_1,Y_2,Y_3)} \underset{\alpha}{\varphi}\right)X - \underset{\alpha}{\varphi}\left(\nabla_{R(Y_1,Y_2,Y_3)} X\right) - S\left(\underset{\alpha}{\varphi} X, R(Y_1, Y_2, Y_3)\right)$$
$$- \underset{\alpha}{\varphi}((\nabla_X R)(Y_1, Y_2, Y_3) + R(\nabla_X Y_1, Y_2, Y_3) + R(Y_1, \nabla_X Y_2, Y_3) + R(Y_1, Y_2, \nabla_X Y_3)$$
$$- \nabla_{R(Y_1,Y_2,Y_3)} X - S(X, R(Y_1, Y_2, Y_3))) + R\left(-\nabla_{\underset{\alpha}{\varphi} X} Y_1 + \underset{\alpha}{\varphi}(\nabla_X Y_1), Y_2, Y_3\right)$$
$$+ R\left(Y_1, -\nabla_{\underset{\alpha}{\varphi} X} Y_2 + \underset{\alpha}{\varphi}(\nabla_X Y_2), Y_3\right) + R\left(Y_1, Y_2, -\nabla_{\underset{\alpha}{\varphi} X} Y_3 + \underset{\alpha}{\varphi}(\nabla_X Y_3)\right)$$
$$= \left(\nabla_{\underset{\alpha}{\varphi} X} R\right)(Y_1, Y_2, Y_3) - \underset{\alpha}{\varphi}(\nabla_X R)(Y_1, Y_2, Y_3),$$

by virtue of $\nabla \underset{\alpha}{\varphi} = 0$. Thus, we have

$$\left(\Phi_{\underset{\alpha}{\varphi}} R\right)(X, Y_1, Y_2, Y_3) = \left(\nabla_{\underset{\alpha}{\varphi} X} R\right)(Y_1, Y_2, Y_3) - \underset{\alpha}{\varphi}(\nabla_X R)(Y_1, Y_2, Y_3). \tag{1.54}$$

For simplicity, let now ∇ be a pure torsion-free Π-connection. Then, using the purity of R and applying the Bianchi's 2nd identity to (1.54), we get

$$\begin{aligned}\left(\Phi_{\underset{\alpha}{\varphi}} R\right)(X, Y_1, Y_2, Y_3) &= \left(\nabla_{\underset{\alpha}{\varphi} X} R\right)(Y_1, Y_2, Y_3) - \underset{\alpha}{\varphi}(\nabla_X R)(Y_1, Y_2, Y_3) \\ &= -(\nabla_{Y_1} R)\left(Y_2, \underset{\alpha}{\varphi} X, Y_3\right) - (\nabla_{Y_2} R)\left(\underset{\alpha}{\varphi} X, Y_1, Y_3\right) \\ &\quad - \underset{\alpha}{\varphi}(\nabla_X R)(Y_1, Y_2, Y_3).\end{aligned}$$

On the other hand, using $\nabla \underset{\alpha}{\varphi} = 0$, we find

$$(\nabla_{Y_2} R)\left(\underset{\alpha}{\varphi} X, Y_1, Y_3\right)$$
$$= \nabla_{Y_2}\left(R\left(\underset{\alpha}{\varphi} X, Y_1, Y_3\right)\right) - R\left(\nabla_{Y_2}\left(\underset{\alpha}{\varphi} X\right), Y_1, Y_3\right) - R\left(\underset{\alpha}{\varphi} X, \nabla_{Y_2} Y_1, Y_3\right)$$

$$
\begin{aligned}
&- R\left(\underset{\alpha}{\varphi} X, Y_1, \nabla_{Y_2} Y_3\right)\\
&= \left(\nabla_{Y_2} \underset{\alpha}{\varphi}\right)(R(X, Y_1, Y_3)) + \underset{\alpha}{\varphi}(\nabla_{Y_2} R(X, Y_1, Y_3)) - R\left(\left(\nabla_{Y_2} \underset{\alpha}{\varphi}\right) X\right)\\
&\quad + \varphi(\nabla_{Y_2} X), Y_1, Y_3) - R\left(\underset{\alpha}{\varphi} X, \nabla_{Y_2} Y_1, Y_3\right) - R\left(\underset{\alpha}{\varphi} X, Y_1, \nabla_{Y_2} Y_3\right)\\
&= \underset{\alpha}{\varphi}\big(\nabla_{Y_2} R(X, Y_1, Y_3)\big) - \underset{\alpha}{\varphi}\big(R\big(\nabla_{Y_2} X, Y_1, Y_3\big)\big) - \underset{\alpha}{\varphi}(R(X, \nabla_{Y_2} Y_1, Y_3))\\
&\quad - \underset{\alpha}{\varphi}(R(X, Y_1, \nabla_{Y_2} Y_3))\\
&= \underset{\alpha}{\varphi}\big(\nabla_{Y_2} R\big)(X, Y_1, Y_3)).
\end{aligned} \tag{1.55}
$$

Similarly, we obtain

$$
(\nabla_{Y_1} R)\left(Y_2, \underset{\alpha}{\varphi} X, Y_3\right) = \underset{\alpha}{\varphi}((\nabla_{Y_1} R)(Y_2, X, Y_3)). \tag{1.56}
$$

Substituting (1.55) and (1.56) into (1.54), and using again the Bianchi's 2nd identity, we obtain

$$
\begin{aligned}
\left(\Phi_{\underset{\alpha}{\varphi}} R\right)(X, Y_1, Y_2, Y_3) &= - \underset{\alpha}{\varphi}((\nabla_{Y_1} R)(Y_2, X, Y_3) - \underset{\alpha}{\varphi}((\nabla_{Y_2} R)(X, Y_1, Y_3)\\
&\quad - \underset{\alpha}{\varphi}((\nabla_X R)(Y_1, Y_2, Y_3))\\
&= - \underset{\alpha}{\varphi}(\sigma\{(\nabla_X R)(Y_1, Y_2)\}, Y_3) = 0,
\end{aligned}
$$

where σ denotes the cyclic sum with respect to X, Y_1 and Y_2. Therefore, by virtue of Theorems 1.9 and 1.21, we have

Theorem 1.23 *The curvature tensor* $\overset{*}{R}$ *of the* $\mathfrak{A}$*-holomorphic connection* $\overset{*}{\nabla}$ *is an* $\mathfrak{A}$*-holomorphic tensor.*

Example 1.1 Let now take $r = 1$, i.e. we consider a $\mathfrak{A}$-holomorphic manifold $X_1(\mathfrak{A}_m)$ of hypercomplex dimension 1. Since $u = v = w = t = 1$, we have from (1.49).

$$
\begin{aligned}
R^i_{jkl} &= R^{1\alpha}_{1\beta 1\gamma 1\delta}\\
&= \left(\partial_{1\beta} \overset{\sigma}{\tau}{}^{1}_{11}\right) B^{\alpha}_{\sigma\gamma\delta} - \left(\partial_{1\gamma} \overset{\sigma}{\tau}{}^{1}_{11}\right) B^{\alpha}_{\sigma\beta\delta} + \left(\overset{\sigma}{\tau}{}^{1}_{11} \overset{\theta}{\tau}{}^{1}_{11} - \overset{\sigma}{\tau}{}^{1}_{11} \overset{\theta}{\tau}{}^{1}_{11}\right) B^{\alpha}_{\sigma\gamma\theta\beta\delta}\\
&= \left(\partial_{1\beta} \overset{\sigma}{\tau}{}^{1}_{11}\right) B^{\alpha}_{\sigma\gamma\delta} - \left(\partial_{1\gamma} \overset{\sigma}{\tau}{}^{1}_{11}\right) B^{\alpha}_{\sigma\beta\delta}\\
&= \left(\partial_\beta \overset{\sigma}{\tau}\right) C^{\alpha}_{\delta\varepsilon} C^{\varepsilon}_{\sigma\gamma} - \left(\partial_\gamma \overset{\sigma}{\tau}\right) C^{\varepsilon}_{\sigma\beta} C^{\alpha}_{\varepsilon\delta}\\
&= \left(C^{\varepsilon}_{\sigma\gamma} \partial_\beta \overset{\sigma}{\tau} - C^{\varepsilon}_{\sigma\beta} \partial_\gamma \overset{\sigma}{\tau}\right) C^{\alpha}_{\delta\varepsilon},
\end{aligned} \tag{1.57}
$$

where $\overset{\sigma}{\tau} = \overset{\sigma}{\tau}{}^{1}_{11}$.

We now assume that

$$C^{\varepsilon}_{\sigma\gamma}\partial_{\beta}\overset{\sigma}{\tau} = C^{\varepsilon}_{\sigma\beta}\partial_{\gamma}\overset{\sigma}{\tau}. \tag{1.58}$$

After contraction with ε^{γ}, from (1.58), we have

$$\partial_{\beta}\overset{\varepsilon}{\tau} = \varepsilon^{\gamma}(\partial_{\gamma}\overset{\sigma}{\tau})C^{\varepsilon}_{\sigma\beta}, \tag{1.59}$$

i.e. by virtue of (1.12) τ is an $\mathfrak{A}$-holomorphic function of $x = x^{\alpha}e_{\alpha} \in \mathfrak{A}_m$.

Conversely, let $\tau = \tau(x)$ be an $\mathfrak{A}$-holomorphic function. Then, from (1.59), we have

$$\begin{aligned} C^{\theta}_{\varepsilon\tau}\partial_{\beta}\overset{\varepsilon}{\tau} &= \varepsilon^{\gamma}\left(\partial_{\gamma}\overset{\sigma}{\tau}\right)C^{\varepsilon}_{\sigma\beta}C^{\theta}_{\varepsilon\tau} \\ &= \varepsilon^{\gamma}\left(\partial_{\gamma}\overset{\sigma}{\tau}\right)C^{\theta}_{\beta\varepsilon}C^{\varepsilon}_{\sigma\tau} \\ &= \varepsilon^{\gamma}\left(\partial_{\gamma}\overset{\sigma}{\tau}\right)C^{\varepsilon}_{\sigma\tau}C^{\theta}_{\beta\varepsilon} \\ &= \varepsilon^{\gamma}\left(\partial_{\sigma}\overset{\varepsilon}{\tau}\right)C^{\sigma}_{\gamma\tau}C^{\theta}_{\beta\varepsilon} \\ &= \left(\partial_{\tau}\overset{\varepsilon}{\tau}\right)C^{\theta}_{\beta\varepsilon}, \end{aligned}$$

i.e. the condition (1.58) is true. On the other hand, the condition (1.58) is equivalent to the Scheffers condition (1.11). Thus, if $\tau = \tau(x)$ is $\mathfrak{A}$-holomorphic, then $R^{i}_{jkl} = 0$. Conversely, if $R = 0$, then from (1.57), we have

$$\begin{aligned} 0 &= R^{1\alpha}_{1\beta1\gamma1\delta}\varepsilon^{\delta} \\ &= \left(C^{\varepsilon}_{\sigma\gamma}\partial_{\beta}\overset{\sigma}{\tau} - C^{\varepsilon}_{\sigma\beta}\partial_{\gamma}\overset{\sigma}{\tau}\right)C^{\alpha}_{\delta\varepsilon}\varepsilon^{\delta} \\ &= C^{\alpha}_{\sigma\gamma}\partial_{\beta}\overset{\sigma}{\tau} - C^{\alpha}_{\sigma\beta}\partial_{\gamma}\overset{\sigma}{\tau}, \end{aligned}$$

i.e. the function $\tau = \tau(x)$ is $\mathfrak{A}$-holomorphic.

Thus, we have.

Theorem 1.24 *Let M_m be a realization of $X_1(\mathfrak{A}_m)$. The connection $\overset{*}{\nabla}$ with components $\overset{*}{\tau} = \overset{\sigma}{\tau}e_{\sigma}$ on $X_1(\mathfrak{A}_m)$ is $\mathfrak{A}$-holomorphic if and only if the real manifold M_m is locally flat.*

Remark 1.4 In particular, if $\overset{\sigma}{\tau} = \overset{\sigma}{\tau}{}^{1}_{11} = \varepsilon^{\sigma}$ $(1 = \varepsilon^{\sigma}e_{\sigma} \in \mathfrak{A}_m)$, then from (1.44), we have $\Gamma^{\alpha}_{\gamma\beta} = C^{\alpha}_{\gamma\beta}$, i.e. M_m is the Vranceanu space [44, 83]. Since $\overset{\sigma}{\tau}e_{\sigma} = \varepsilon^{\sigma}e_{\sigma} = 1$ is holomorphic, we see that the Vranceanu space is locally flat.

2 Anti-Hermitian Geometry

In this chapter, we study the pseudo-Riemannian metric on holomorphic manifolds. In Sect. 2.1, we give the condition for a hypercomplex anti-Hermitian metric to be holomorphic; also, we prove that there exists a one-to-one correspondence between hypercomplex anti-Kähler manifolds and anti-Hermitian manifolds with an $\mathfrak{A}$-holomorphic metrics. In Sect. 2.2, we discuss complex Norden manifolds. We define the twin Norden metric; the main theorem of this section is that the Levi–Civita connection of Kähler-Norden metric coincides with the Levi–Civita connection of twin Norden metric. In Sect. 2.3, we consider Norden-Hessian structures. We give the condition for a Norden-Hessian manifold to be Kähler. Section 2.4 is devoted to the analysis of twin Norden metric connections with torsion. In Sects. 2.5–2.10, we focus our attention to pseudo-Riemannian 4-manifolds of neutral signature. The main purpose of these sections is to study complex Norden metrics on 4-dimensional Walker manifolds. We discuss the integrability and Kahler (holomorphic) conditions for these structures. The curvature properties for Norden-Walker metrics are also investigated, and examples of Norden-Walker metrics are constructed from an arbitrary harmonic function of two variables. We define the isotropic Kähler structures and, moreover, show that a proper almost complex structure on an almost Norden-Walker manifold is isotropic Kähler. We also consider the quasi-Kähler-Norden metric and give the condition for an almost Norden manifold to be quasi-Kähler-Norden. Finally, we give progress to the conjecture of Goldberg under the additional restriction on Norden-Walker metric.

2.1 Equivalence of Holomorphic and Anti-Kähler Conditions

Let M_{mr} be a pseudo-Riemannian manifold with metric g, and let on M_{mr} be given the regular hypercomplex Π-structure:

A. Salimov, *Applications of Holomorphic Functions in Geometry*,
Frontiers in Mathematics, https://doi.org/10.1007/978-981-99-1296-4_2

$$\Pi = \left\{ \underset{\sigma}{\varphi}{}^{i}_{j} \right\}, \underset{\sigma}{\varphi}{}^{i}_{j} = \delta^u_v C^\alpha_{\sigma\beta}, i = u\alpha, j = v\beta;$$

$$i, j = 1, \ldots, mr; \alpha, \beta, \sigma = 1, \ldots, m; u, v = 1, \ldots, r$$

An *anti-Hermitian metric* with respect to the Π-structure is a pseudo-Riemannian metric g such that

$$g\left(\underset{\alpha}{\varphi} X, Y\right) = gX, \underset{\alpha}{\varphi} Y, \alpha = 1, ..., m, \tag{2.1}$$

for any $X, Y \in \mathfrak{I}^1_0(M_{mr})$, i.e. g is pure with respect to the regular hypercomplex Π-structure. Such metrics were studied in [79], where they were said to be B-metrics, since the metric tensor g with respect to the Π-structure is B-tensor according to the terminology accepted by Norden [40]. If (M_{mr}, Π) is an almost hypercomplex manifold with anti-Hermitian metric g, we say that the triple (M_{mr}, Π, g) is an *almost anti-Hermitian manifold*. If the Π-structure is integrable, we say that the triple (M_{mr}, Π, g) is an *anti-Hermitian manifold.*

An anti-Hermitian metric g is called an *almost* $\mathfrak{A}$*-holomorphic metric* (see Remark 1.3) if

$$\begin{aligned}\left(\Phi_{\underset{\alpha}{\varphi}} g\right)(X, Y, Z) &= (\underset{\alpha}{\varphi} X)(g(Y, Z)) - X\left(g\left(\underset{\alpha}{\varphi} Y, Z\right)\right) + g\left(\left(L_Y \underset{\alpha}{\varphi}\right)X, Z\right) \\ &\quad + g\left(Y, \left(L_Z \underset{\alpha}{\varphi}\right)X\right) \\ &= \left(L_{\underset{\alpha}{\varphi} X} g - L_X\left(g \circ \underset{\alpha}{\varphi}\right)\right)(Y, Z) = 0, \ \alpha = 1, ..., m, \end{aligned} \tag{2.2}$$

for any $X, Y, Z \in \mathfrak{I}^1_0(M_{mr})$, where $\Phi_{\underset{\alpha}{\varphi}}$, $\alpha = 1, ..., m$ are Tachibana operators applied to an anti-Hermitian metric (see Sect. 1.6). If (M_{mr}, Π, g) is an almost anti-Hermitian manifold with almost $\mathfrak{A}$-holomorphic metric g, we say that (M_{mr}, Π, g) is an *almost* $\mathfrak{A}$*-holomorphic anti-Hermitian manifold*. If the Π-structure is integrable, then we say that the triple (M_{mr}, Π, g) is an $\mathfrak{A}$*-holomorphic anti-Hermitian manifold.*

Theorem 2.1 *An almost anti-Hermitian manifold* (M_{mr}, Π, g) *is an almost* $\mathfrak{A}$*-holomorphic if and only if* $\nabla \underset{\alpha}{\varphi} = 0$, $\alpha = 1, ..., m$, *where* ∇ *is the Levi–Civita connection of* g.

Proof Using (2.1) and $L_X Y = [X, Y] = \nabla_X Y - \nabla_Y X$, from (2.2), we get

$$\begin{aligned}&\left(\Phi_{\underset{\alpha}{\varphi}} g\right)(X, Z_1, Z_2) \\ &= \left(L_{\underset{\alpha}{\varphi} X} g - L_X\left(g \circ \underset{\alpha}{\varphi}\right)\right)(Z_1, Z_2) + g\left(Z_1, \underset{\alpha}{\varphi} L_X Z_2\right) - g\left(\underset{\alpha}{\varphi} Z_1, L_X Z_2\right)\end{aligned}$$

$$
\begin{aligned}
&= \left(\underset{\alpha}{\varphi} X\right) g(Z_1, Z_2) - Xg\left(\underset{\alpha}{\varphi} Z_1, Z_2\right) - g\left(\nabla_{\underset{\alpha}{\varphi} X} Z_1, Z_2\right) + g\left(\nabla_{Z_1} \underset{\alpha}{\varphi} X, Z_2\right) \\
&\quad - g\left(Z_1, \nabla_{\underset{\alpha}{\varphi} X} Z_2\right) + g\left(Z_1, \nabla_{Z_2} \underset{\alpha}{\varphi} X\right) + g\left(\underset{\alpha}{\varphi}(\nabla_X Z_1), Z_2\right) - g\left(\underset{\alpha}{\varphi}(\nabla_{Z_1} X), Z_2\right) \\
&\quad + g\left(\underset{\alpha}{\varphi} Z_1, \nabla_X Z_2\right) - g\left(Z_1, \underset{\alpha}{\varphi}(\nabla_{Z_2} X)\right)
\end{aligned}
\tag{2.3}
$$

We find

$$
\begin{aligned}
& g\left(\nabla_{Z_1} \underset{\alpha}{\varphi} X, Z_2\right) - g\left(\underset{\alpha}{\varphi}(\nabla_{Z_1} X), Z_2\right) \\
&\quad + g\left(Z_1, \nabla_{Z_2} \underset{\alpha}{\varphi} X\right) - g\left(Z_1, \underset{\alpha}{\varphi}(\nabla_{Z_2} X)\right) \\
&= g((\nabla\varphi)(X, Z_1), Z_2) + g(Z_1, (\nabla\varphi)(X, Z_2)).
\end{aligned}
\tag{2.4}
$$

Substituting (2.4) into (2.3), we have

$$
\begin{aligned}
&\left(\Phi_{\underset{\alpha}{\varphi}} g\right)(X, Z_1, Z_2) \\
&= \left(\underset{\alpha}{\varphi} X\right) g(Z_1, Z_2) - Xg\left(\underset{\alpha}{\varphi} Z_1, Z_2\right) + g\left(\left(\nabla \underset{\alpha}{\varphi}\right)(X, Z_1), Z_2\right) \\
&\quad g\left(Z_1, \left(\nabla \underset{\alpha}{\varphi}\right)(X, Z_2)\right) - g\left(\nabla_{\underset{\alpha}{\varphi} X} Z_1, Z_2\right) - g\left(Z_1, \nabla_{\underset{\alpha}{\varphi} X} Z_2\right) \\
&\quad + g\left(\underset{\alpha}{\varphi}(\nabla_X Z_1), Z_2\right) + g\left(\underset{\alpha}{\varphi} Z_1, \nabla_X Z_2\right)
\end{aligned}
\tag{2.5}
$$

On the other hand, with respect to the Levi–Civita connection ∇, we have

$$
\begin{aligned}
&\left(\underset{\alpha}{\varphi} X\right) g(Z_1, Z_2) - g\left(\nabla_{\underset{\alpha}{\varphi} X} Z_1, Z_2\right) \\
&\quad - g\left(Z_1, \nabla_{\underset{\alpha}{\varphi} X} Z_2\right) = \left(\nabla_{\underset{\alpha}{\varphi} X} g\right)(Z_1, Z_2) = 0
\end{aligned}
\tag{2.6}
$$

and

$$
\begin{aligned}
&- Xg\left(\underset{\alpha}{\varphi} Z_1, Z_2\right) + g\left(\underset{\alpha}{\varphi}(\nabla_X Z_1), Z_2\right) + g\left(\underset{\alpha}{\varphi} Z_1, \nabla_X Z_2\right) \\
&= -g\left(\left(\nabla_X \underset{\alpha}{\varphi}\right) Z_1, Z_2\right)
\end{aligned}
\tag{2.7}
$$

By virtue of (2.6) and (2.7), the operator (2.5) reduces to

$$
\left(\Phi_{\underset{\alpha}{\varphi}} g\right)(X, Z_1, Z_2) = -g\left(\left(\nabla_X \underset{\alpha}{\varphi}\right) Z_1, Z_2\right)
$$

$$+g\left(\left(\nabla_{Z_1}\underset{\alpha}{\varphi}\right)X, Z_2\right)+g\left(Z_1,\left(\nabla_{Z_2}\underset{\alpha}{\varphi}\right)X\right), \tag{2.8}$$

for any $X, Z_1, Z_2 \in \Im_0^1(M_{mr})$. From (2.8), we easily see that if $\nabla \underset{\alpha}{\varphi} = 0$, then $\Phi_{\underset{\alpha}{\varphi}} g = 0$.

Conversely, let now $\Phi_{\underset{\alpha}{\varphi}} g = 0$. Then, similarly to (2.8), we have

$$\left(\Phi_{\underset{\alpha}{\varphi}} g\right)(Z_2, Z_1, X) = -g\left(\left(\nabla_{Z_2}\underset{\alpha}{\varphi}\right)Z_1, X\right)+g\left(\left(\nabla_{Z_1}\underset{\alpha}{\varphi}\right)Z_2, X\right)+g\left(Z_1,\left(\nabla_X\underset{\alpha}{\varphi}\right)Z_2\right) \tag{2.9}$$

Using $g\left(Z,\left(\nabla_Y\underset{\alpha}{\varphi}\right)X\right) = g\left(\left(\nabla_Y\underset{\alpha}{\varphi}\right)Z, X\right)$, from the sum of (2.8) and (2.9), we find

$$\left(\Phi_{\underset{\alpha}{\varphi}} g\right)(X, Z_1, Z_2)+\left(\Phi_{\underset{\alpha}{\varphi}} g\right)(Z_2, Z_1, X) = 2g\left(X,\left(\nabla_{Z_2}\underset{\alpha}{\varphi}\right)Z_2\right), \tag{2.10}$$

for any $X, Z_1, Z_2 \in \Im_0^1(M_{mr})$. Now, putting $\Phi_{\underset{\alpha}{\varphi}} g = 0$ in (2.10), we find $\nabla \underset{\alpha}{\varphi} = 0, \alpha = 1, ..., m$. Thus, the proof of Theorem 2.1 is completed.

Using the equivalence between the integrability and the almost integrability conditions for regular Π-structures [27], from Theorem 2.1, we obtain the following useful integrability condition:

Theorem 2.2 *The regular Π-structure on almost anti-Hermitian manifold is integrable if g is an almost $\mathfrak{A}$-holomorphic metric, i.e.*

$$\Phi_{\underset{\alpha}{\varphi}} g = 0, \alpha = 1, ..., m.$$

From Theorem 2.2, it follows that the almost anti-Hermitian manifold with an almost $\mathfrak{A}$-holomorphic metric has an integrable Π-structure, and therefore, the manifold of this type is an $\mathfrak{A}$-holomorphic anti-Hermitian manifold.

Also, from Theorem 1.19 and Theorem 2.1, we have

Theorem 2.3 *The Levi–Civita connection ∇ of $\mathfrak{A}$-holomorphic anti-Hermitian manifolds is a pure connection.*

An *anti-Kähler manifold* can be defined as a triple (M_{mr}, Π, g) which consists of a manifold M_{mr} endowed with a hypercomplex integrable regular Π-structure and a pseudo-Riemannian metric g such that $\nabla \underset{\alpha}{\varphi} = 0$, $\alpha = 1, ..., m$, where ∇ is the Levi–Civita connection of g and the metric g is assumed to be Hermitian: $g\left(\underset{\alpha}{\varphi} X, Y\right) =$

$g\left(X, \underset{\alpha}{\varphi} Y\right), \alpha = 1, \ldots, m$. Thus from Theorem 2.1, we see that there exists a one-to-one correspondence between anti-Kähler manifolds and $\mathfrak{A}$-holomorphic anti-Hermitian manifolds.

From (2.1), we have

$$g_{ij} = g_{u\alpha v\beta} = G_{uv\sigma} C^{\sigma}_{\alpha\beta} \tag{2.11}$$

for arbitrary functions $G_{uv\sigma}$ (see Sect. 1.5). The corresponding hypercomplex tensor $\overset{*}{g}_{uv}$ is defined by

$$\overset{*}{g}_{uv} = g_{u\alpha v\beta} \varepsilon^{\beta} e^{\alpha} = G_{uv\alpha} e^{\alpha}, \tag{2.12}$$

where $e^{\alpha} = \varphi^{\alpha\beta} e_{\beta}$ and $\varphi^{\alpha\beta}$ is the Frobenius metric. Since $\overset{*}{g}_{uv}$ is symmetric, nonsingular $\left(\mathrm{Det}\left(\overset{*}{g}_{uv}\right) \neq 0\right)$ and $\mathfrak{A}$-holomorphic, it follows that $d\overset{*}{s}{}^{2} = \underset{uv}{\overset{*}{g}} dz^{u} dz^{v}$ can be regarded as hypercomplex anti-Kähler metric in $X_r(\mathfrak{A})$.

Since the Levi–Civita connection ∇ of anti-Kähler manifolds is a pure connection, by virtue of (1.44), we obtain

$$K^{i}_{kj} = K^{u\alpha}_{w\gamma v\beta} = \overset{\sigma}{\mathrm{k}}{}^{u}_{wv} C^{\mu}_{\sigma\gamma} C^{\alpha}_{\mu\beta} = \overset{\sigma}{\mathrm{k}}{}^{u}_{wv} B^{\alpha}_{\sigma\gamma\beta}, \overset{\sigma}{\mathrm{k}}{}^{u}_{wv} = \overset{\sigma}{\mathrm{k}}{}^{v}_{wv}, \tag{2.13}$$

where K^{i}_{kj} are components of ∇ and $i = u\alpha, j = v\beta, k = w\gamma$. Substituting (2.11) and (2.13) into $\nabla_k g_{ij} = \partial_k g_{ij} - K^{m}_{ki} g_{mj} - K^{m}_{kj} g_{im} = 0,\ m = tv$, we find

$$\partial_{wv}\left(G_{uv\sigma} C^{\sigma}_{\alpha\beta}\right) = \overset{\sigma}{\mathrm{k}}{}^{t}_{wu} C^{v}_{\sigma\gamma} C^{\mu}_{v\alpha} G_{tv\tau} C^{\tau}_{\mu\beta} + \overset{\sigma}{\mathrm{k}}{}^{t}_{wv} C^{v}_{\sigma\gamma} C^{\mu}_{v\beta} G_{tu\tau} C^{\tau}_{\mu\alpha}$$

After contraction with $\varepsilon^{\beta}\varepsilon^{\gamma}e^{\alpha}$, by virtue of (1.5) and (1.9), we obtain

$$\varepsilon^{\gamma} \partial_{wv} G_{uv\alpha} e^{\alpha} = \left(\overset{\sigma}{\mathrm{k}}{}^{t}_{wu} G_{tv\mu} + \mathrm{k}^{\sigma}_{wv} G_{tu\mu}\right) e_{\sigma} e^{\mu}$$

By virtue of (1.14), (1.45) and (2.12), we have

$$\frac{\partial \overset{*}{g}_{uv}}{\partial z^{w}} = \overset{*}{K}{}^{t}_{wu} \overset{*}{g}_{tv} + \overset{*}{K}{}^{t}_{wv} \overset{*}{g}_{ut},$$

from which we easily see that the Christoffel symbol $\overset{*}{K}$ has components

$$\overset{*}{K}{}^{u}_{wv} = \frac{1}{2} \overset{*}{g}{}^{ut} \left(\frac{\partial \overset{*}{g}_{tv}}{\partial z^{w}} + \frac{\partial \overset{*}{g}_{wt}}{\partial z^{v}} - \frac{\partial \overset{*}{g}_{wv}}{\partial z^{t}} \right). \tag{2.14}$$

Since the hypercomplex anti-Kähler metric $\overset{*}{g}$ is $\mathfrak{A}$-holomorphic, there exist the successive derivatives of $\overset{*}{g}$ by virtue of Remark 1.1, i.e. from (2.14), it follows that $\overset{*}{K}{}_{wv}^{\;\;u}$ is $\mathfrak{A}$-holomorphic.

Thus, we have

Theorem 2.4 *The Levi–Civita connection of an anti-Kähler manifold is $\mathfrak{A}$-holomorphic.*

Since an anti-Kähler manifold of hypercomplex dimension 1 (i.e. $r = 1$) is flat (see Theorem 1.24 and Theorem 2.4), we assume in the sequel that $mr \geq 2m$, i.e. $r \geq 2$.

Using Theorem 1.21, we have

Theorem 2.5 *The Riemannian curvature tensor of an anti-Kähler manifold is pure.*

From Theorem 1.23, we have

Theorem 2.6 *The Riemannian curvature tensor of an anti-Kähler manifold is $\mathfrak{A}$-holomorphic.*

From Theorem 1.11, we have: a necessary and sufficient condition for an exact 1-form df, $f \in \Im_0^0(M_{mr})$ to be $\mathfrak{A}$-holomorphic is that the associated 1-forms $df \circ \underset{\alpha}{\varphi}$, $\alpha = 1, ..., m$ be closed, i.e.

$$d\left(df \circ \underset{\alpha}{\varphi}\right) = 0, \alpha = 1, ..., m. \tag{2.15}$$

If there exist some functions $\underset{\alpha}{g}$, $\alpha = 1, ..., m$, in a hypercomplex anti-Kähler manifold such that $df \circ \underset{\alpha}{\varphi} = d\underset{\alpha}{g}$, $\alpha = 1, ..., m$, for a function f, then we shall call f a *$\mathfrak{A}$-holomorphic function* and $\underset{\alpha}{g}$, $\alpha = 1, ..., m$, associated functions. We notice that Eq. (2.15) is equivalent to $df \circ \underset{\alpha}{\varphi} = d\underset{\alpha}{g}$, $\alpha = 1, ..., m$, only locally. Hence, the condition for f to be locally $\mathfrak{A}$-holomorphic also is given by.

$$\underset{\alpha}{\varphi}{}_i^m \partial_m f = \partial_i \underset{\alpha}{g}.$$

Let now (M_{mr}, Π, g) be a hypercomplex anti-Kähler manifold. Then, using Theorems 1.21 and 2.1, by virtue of (1.54), we find that in these manifolds, the covariant derivative of the curvature tensor ∇R is also pure. Therefore, the covariant derivative of the Ricci tensor $R_{ji} = R_{sji}^{s} = g^{ts} R_{tjis}$ is pure in all its indices, and hence,

$$\underset{\alpha}{\varphi}{}_t^s \nabla_s R_{ji} = \underset{\alpha}{\varphi}{}_j^s \nabla_t R_{si}, \alpha = 1, \ldots, m.$$

Contracting this equation with g^{ji}, we find

$$\begin{aligned}\underset{\alpha}{\varphi}{}_{t}^{s}\nabla_s r &= g^{ji}\underset{\alpha}{\varphi}{}_{j}^{s}\nabla_t R_{si} = \nabla_t\left(g^{ji}\underset{\alpha}{\varphi}{}_{j}^{s}R_{si}\right)\\ &= \nabla_t \underset{\alpha}{\overset{*}{r}} \Leftrightarrow \varphi\underset{\alpha}{\varphi}{}_{t}^{s}\partial_s r = \partial_t \underset{\alpha}{\overset{*}{r}}, \alpha = 1, \ldots, m,\end{aligned} \tag{2.16}$$

where $r = g^{ij}R_{ij}$ is the scalar curvature and $\underset{\alpha}{\overset{*}{r}} = g^{ji}\underset{\alpha}{\varphi}{}_{j}^{s}R_{si}$.

From (2.16),we have

Theorem 2.7 *The scalar curvature r of an anti-Kähler manifold is a locally* $\mathfrak{A}$*-holomorphic function.*

2.2 Kähler-Norden Manifolds

If $\mathfrak{A}_m = \mathbb{C}$ $(m = 2)$ is a complex algebra and $\Pi = \{I, \varphi\}$, $\varphi^2 = -I$, $I = id_{M_{mr}}$, then the anti-Hermitian manifold (M_{2r}, φ, g) is called an *almost Norden manifold*; also, if the Π-structure is integrable, the triple (M_{2r}, φ, g) is called a *Norden manifold* [20]. It is important that the Norden metric g is necessarily neutral metric. Metrics of this kind have been also studied under the names of pure and B-metrics (see, for example, [6–8, 16–18, 20, 22, 23, 25, 28, 31, 32, 38, 39, 42, 43, 45, 46, 51–59, 63–67, 76, 77, 81]).

Let now (M_{2r}, φ, g) be an almost Norden manifold. A *twin Norden metric* of almost Norden manifold is defined by

$$G(X, Y) = (g \circ \varphi)(X, Y) = g(\varphi X, Y) = g(X, \varphi Y), \tag{2.17}$$

for any $X, Y \in \mathfrak{I}_0^1(M_{2r})$. One can easily prove that G is a pure metric, i.e. Det$G \neq 0$ and

$$\begin{aligned}G(\varphi X, Y) &= (g \circ \varphi)(\varphi X, Y) = g(\varphi(\varphi X), Y) = g(\varphi X, \varphi Y)\\ &= (g \circ \varphi)(X, \varphi Y) = G(X, \varphi Y)\end{aligned}$$

The twin metric G is also called the associated (or dual) metric of g, and it plays a role similar to the Kähler form in Hermitian geometry. We shall now apply the Tachibana operator to the pure Riemannian metric G:

$$\begin{aligned}(\Phi_\varphi G)(X, Y, Z) &= (L_{\varphi X}G - L_X(G \circ \varphi))(Y, Z) + G(Y, \varphi L_X Z) - G(\varphi Y, L_X Z)\\ &= (\Phi_\varphi g)(X, \varphi Y, Z) + g(N_\varphi(X, Y), Z)\end{aligned} \tag{2.18}$$

where N_φ is the Nijenhuis tensor defined by

$$N_\varphi(X, Y) = [\varphi X, \varphi X] - \varphi[X, \varphi Y] - \varphi[\varphi X, Y] - [X, Y].$$

It is clear that if ∇ is a torsion-free connection, then

$$N_\varphi(X,Y) = \left(\nabla_{\varphi X}\varphi\right)Y - \left(\nabla_{\varphi Y}\varphi\right)X - \varphi((\nabla_X\varphi)Y - (\nabla_Y\varphi)X).$$

Since $\Phi_\varphi g = 0 \Leftrightarrow \nabla\varphi = 0 \Rightarrow N_\varphi = 0$ and $\Phi_\varphi G = 0 \Leftrightarrow \nabla\varphi = 0 \Rightarrow N_\varphi = 0$ (see Theorem 2.1), from (2.18), we have.

Theorem 2.8 *Let (M_{2r}, φ, g) be an almost Norden manifold and G be its twin metric. The following conditions are equivalent:*

(a) $\Phi_\varphi g = 0$,
(b) $\Phi_\varphi G = 0$.

Let now (M_{2r}, φ, g) be an anti-Kähler (or a Kähler-Norden) manifold. We denote by ∇_g the covariant differentiation with respect to the Levi–Civita connection of g. Then we have

$$\nabla_g G = \left(\nabla_g g\right)\circ\varphi + g\circ\left(\nabla_g\varphi\right) = g\circ\left(\nabla_g\varphi\right),$$

which implies $\nabla_g G = 0$ by virtue of Theorem 2.1. Therefore, we have

Theorem 2.9 *Let (M_{2r}, φ, g) be a Kähler-Norden manifold. Then, the Levi–Civita connection of Norden metric g coincides with the Levi–Civita connection of twin Norden metric G.*

Now, using Theorem 2.9, we will present the alternative proof of Theorem 2.5.

Let R and S be the curvature tensors of g and G, respectively. Then, for a Kähler-Norden manifold, we have $R = S$ by means of the Theorem 2.9. Applying Ricci's identity to φ:

$$\begin{aligned}\nabla_X((\nabla_Y\varphi)Z) - \nabla_Y((\nabla_X\varphi)Z) = R(X,Y)\varphi Z - \varphi(R(X,Y)Z) + \left(\nabla_{[X,Y]}\varphi\right)Z \\ + (\nabla_Y\varphi)(\nabla_X Z) - (\nabla_X\varphi)(\nabla_Y Z),\end{aligned}$$

we get

$$\varphi(R(X,Y)Z) = R(X,Y)\varphi Z \tag{2.19}$$

by virtue of $\nabla\varphi = 0$. Hence, from (2.19) and $R(X_1, X_2, X_3, X_4) = g(R(X_1, X_2)X_3, X_4)$, we find that R is pure with respect to X_3 and X_4:

$$R(X_1, X_2, \varphi X_3, X_4) = g(R(X_1, X_2)\varphi X_3, X_4)$$

$$
\begin{aligned}
&= g(\varphi(R(X_1, X_2)X_3), X_4) \\
&= g(R(X_1, X_2)X_3, \varphi X_4) \\
&= R(X_1, X_2, X_3, \varphi X_4)
\end{aligned}
$$

On the other hand, since S being the curvature tensor formed by twin metric G and $S(X_1, X_2, X_3, X_4) = G(S(X_1, X_2)X_3, X_4)$, we have

$$S(X_1, X_2, X_3, X_4) = S(X_3, X_4, X_1, X_2). \tag{2.20}$$

Taking account of (2.17), (2.19) and $R = S$, we find that

$$
\begin{aligned}
S(X_1, X_2, X_3, X_4) &= G(S(X_1, X_2)X_3, X_4) \\
&= g(\varphi(S(X_1, X_2)X_3), X_4) \\
&= g(S(X_1, X_2)X_3, \varphi X_4) \\
&= g(R(X_1, X_2)X_3, \varphi X_4) \\
&= R(X_1, X_2, X_3, \varphi X_4)
\end{aligned}
$$

and

$$
\begin{aligned}
S(X_3, X_4, X_1, X_2) &= G(S(X_3, X_4)X_1, X_2) \\
&= g(\varphi(S(X_3, X_4)X_1), X_2) \\
&= g(S(X_3, X_4)X_1, \varphi X_2) \\
&= g(R(X_3, X_4)X_1, \varphi X_2) \\
&= R(X_3, X_4, X_1, \varphi X_2) \\
&= R(X_1, \varphi X_2, X_3, X_4).
\end{aligned}
$$

Thus, Eq. (2.20) becomes.

$$R(X_1, X_2, X_3, \varphi X_4) = R(X_1, \varphi X_2, X_3, X_4),$$

which shows that $R = R(X_1, X_2, X_3, X_4)$ is pure with respect to X_2 and X_4. Since $R(X_1, X_2, X_3, X_4) = R(X_3, X_4, X_1, X_2)$, therefore $R = R(X_1, X_2, X_3, X_4)$ is a pure tensor with respect to all arguments.

Also, from (2.16), we see that the curvature scalar r of Kähler-Norden manifold is a locally $\mathbb{C}$-holomorphic function, and the associated function $\overset{*}{r}$ coincides with the scalar curvature of the twin metric G.

2.3 Hessian-Norden Structures

Let (M_{2r}, g) be a Riemannian manifold with a metric tensor g. The *gradient of a function* $f \in \mathfrak{I}_0^0(M_{2r})$ is the vector field metrically equivalent to the differential $df \in \mathfrak{I}_1^0(M_{2r})$[41]. In terms of a coordinate system,

$$\nabla f = \left(g^{ij}\partial_i f\right)\partial_j$$

Thus,

$$g(\nabla f, X) = g^{ij}(\partial_i f)X^k g_{jk} = Xf = (df)(X).$$

The *Hessian of a function* $f \in \mathfrak{I}_0^0(M_{2r})$ is its second covariant differential $h = \nabla(\nabla f) = \nabla^2 f$ with respect to the Levi–Civita connection ∇ of g,

i.e. $h \in \mathfrak{I}_2^0(M_{2r})$. Since $\nabla_Y f = Yf = (df)(Y)$ and $\nabla_X Y - \nabla_Y X - [X, Y] = 0$, we have

$$\begin{aligned}
h(Y, X) &= (\nabla(\nabla f))(Y, X) = (\nabla(df))(Y, X) \\
&= X((df)(Y)) - (df)(\nabla_X Y) = XYf - (\nabla_X Y)f, \\
h(X, Y) &= (\nabla(\nabla f))(X, Y) = (\nabla(df))(X, Y) \\
&= Y((df)(X) - (df)(\nabla_Y X) = YXf - (\nabla_Y X)f, \\
h(Y, X) - h(Y, X) &= XYf - (\nabla_X Y)f - (YXf - (\nabla_Y X)f) \\
&= [X, Y]f - (\nabla_X Y - \nabla_Y X)f \\
&= [X, Y]f - [X, Y]f = 0,
\end{aligned}$$

i.e. h is a symmetric tensor field. Also, we see that

$$\begin{aligned}
g(\nabla_X(\nabla f), Y) &= g(\nabla_X\left(\left(g^{ij}\partial_i f\right)\partial_j\right), Y) = g\Big(X^s\left(\partial_s g^{ij}\right)(\partial_i f)\partial_j \\
&+ g^{ij}X^s(\partial_s\partial_i f)\partial_j + g^{ij}(\partial_i f)X^s\Gamma_{sj}^m\partial_m, Y^k\partial_k\Big) = -X^s g^{ij}\left(\partial_s g_{jk}\right)(\partial_i f)Y^k \\
&+ g\left(g^{ij}X^s(\partial_s\partial_i f)\partial_j + g^{ij}(\partial_i f)X^s\Gamma_{sj}^m\partial_m, Y^k\partial_k\right) = \left(-\Gamma_{sj}^l g_{lk} - \Gamma_{sk}^l g_{jl}\right)X^s g^{ij}(\partial_i f)Y^k \\
&+ X^s Y^k(\partial_s\partial_i f)g^{ij}g_{jk} + \Gamma_{sj}^m(\partial_i f)X^s Y^k g^{ij}g_{mk} = X^s Y^k(\partial_s\partial_k f) - X^s\Gamma_{sk}^l\delta_l^i(\partial_i f)Y^k \\
&= X^s\partial_s\left(Y^k\partial_k f\right) - X^s\left(\partial_s Y^k\right)\partial_k f - X^s\Gamma_{sk}^l(\partial_l f)Y^k = X^s\partial_s(Yf) \\
&- X^s\left(\partial_s Y^l + \Gamma_{sk}^l Y^k\right)\partial_l f = XYf - (\nabla_X Y)f,
\end{aligned}$$

i.e.

$$h(X, Y) = g(\nabla_X(\nabla f), Y).$$

If $\mathrm{Det}\, h \neq 0$, then $h = \nabla^2 f$ defines a new indefinite metric on M_{2r} and is called a *Hessian metric*. Let Γ_{ij}^p be the Christoffel symbol and $R_{ijk}{}^m$ be the components of the curvature tensor fields produced by the Riemannian metric g. If h^{pk} are the contravariant components of the pseudo-Riemannian Hessian metric h, then the components of Levi–Civita connection ${}^h\widetilde{\nabla}$ of h are given by the following formula:

$$ {}^h\widetilde{\Gamma}_{ij}^p = \Gamma_{ij}^p + \frac{1}{2} h^{pk}\Big[\big(\nabla_i \nabla_j \nabla_k f\big) + \Big(R_{ikj}{}^m + R_{jki}{}^m\Big)\nabla_m f\Big]. $$

Let (M_{2r}, g, φ) be a Kähler-Norden manifold. If there exists a function $\overset{*}{f}$ on M_{2r} such that $df \circ \varphi = d\overset{*}{f}$ for a function f, then we shall call f a *holomorphic function* and $\overset{*}{f}$ its *associated function* (see Sect. 2.1). If such a function f is defined locally, then we call it a *locally holomorphic function*.

Remark 2.1 If (M_{2r}, φ) is a complex manifold, then in terms of a real coordinates $\left(x^i, x^{\bar{i}}\right)$, $i = 1, ..., r; \bar{i} = r+1, ..., 2r$, the equation $df \circ \varphi = d\overset{*}{f}$ reduces to

$$ \begin{cases} \partial_{\bar{i}} f = \partial_i \overset{*}{f}, \\ \partial_i f = -\partial_{\bar{i}} \overset{*}{f}, \end{cases} $$

which is the Cauchy-Riemann equations for the complex function $F = \overset{*}{f} + if$ (see [26, p. 122]).

We notice that the condition for f to be locally holomorphic is given also by (see Theorem 1.11).

$$ \left(\Phi_\varphi df\right)_{ij} = \varphi_i^m \partial_m \partial_j f - \partial_i\left(\varphi_j^m \partial_m f\right) + \left(\partial_j \varphi_i^m\right)\partial_m f = 0. $$

If we assume that f is holomorphic, then, from (1.34), we have

$$ \begin{aligned} (\Phi_\varphi(df))(X,Y) &= (\varphi X)((df)(Y)) - X((df)(\varphi Y)) + (df)((L_Y\varphi)(X)) \\ &= \varphi(X)((df)(Y)) - X((df)(\varphi Y)) + (df)([Y, \varphi X] - \varphi([Y,X])) \\ &= \varphi(X)((df)(Y)) - X((df)(\varphi Y)) \\ &\quad + (df)(\nabla_Y \varphi X - \nabla_{\varphi X} Y - \varphi(\nabla_Y X - \nabla_X Y)) \\ &= (\nabla_{\varphi X} df)(Y) - (\nabla_X df)(\varphi Y) + (df)(\nabla_{\varphi X} Y) - (df)(\nabla_X \varphi Y) \\ &\quad + (df)((\nabla\varphi)(X,Y) - \nabla_{\varphi X} Y + \varphi(\nabla_X Y)) \\ &= (\nabla_{\varphi X} df)(Y) - (\nabla_X df)(\varphi Y) - (\nabla\varphi)(Y,X) = 0 \end{aligned} \tag{2.21} $$

We now consider a holomorphic function f on a Kähler-Norden manifold (M_{2r}, g, φ). On a Kähler-Norden manifold (M_{2r}, g, φ) $(\nabla\varphi = 0)$, Eq. (2.21) is equivalent to the equation.

$$(\nabla^2 f)(Y, \varphi X) = (\nabla^2 f)(\varphi Y, X),$$

i.e. $h = \nabla^2 f$ is pure, and a manifold $(M_{2r}, \varphi, h = \nabla^2 f)$ is a Norden manifold. Thus, h naturally defines a new Norden metric on Kähler-Norden manifold (M_{2r}, g, φ). We call it *Hessian-Norden metric*. Thus, we have.

Theorem 2.10 *Let (M_{2r}, g, φ) be a Kähler-Norden manifold. Then M_{2r} admits a Hessian-Norden structure $(\varphi, h = \nabla^2 f)$, if $f \in \Im_0^0(M_{2r})$ is holomorphic.*

Let (M_{2r}, g, φ) be a Kähler-Norden manifold. Then g is holomorphic, and the curvature tensor R of g is pure with respect to the structure φ (Theorem 2.5). Let $(M_{2n}, h = \nabla^2 f, \varphi)$ be a Hessian-Norden structure which exists on a Kähler-Norden manifold. Then

$$(\nabla^2 f)(\varphi X, Y) = (\nabla^2 f)(X, \varphi Y),$$

from which we have

$$(\nabla^3 f)(\varphi X, Y, Z) = (\nabla^3 f)(X, \varphi Y, Z).$$

Using Ricci equation for $(df)X = Xf = \nabla_X f$, from here we obtain

$$\begin{aligned}
&(\nabla^3 f)(X, \varphi Y, Z) = \nabla_Z(\nabla_{\varphi Y}(\nabla_X f)) \\
&= \nabla_{\varphi Y}(\nabla_Z(\nabla_X f)) - (df)(R(Z, \varphi Y)X) \\
&= (\nabla^3 f)(X, Z, \varphi Y) - (df)(R(Z, \varphi Y)X)
\end{aligned} \tag{2.22}$$

and

$$\begin{aligned}
&(\nabla^3 f)(\varphi X, Y, Z) = \nabla_Z(\nabla_Y(\nabla_{\varphi X} f)) \\
&= \nabla_Y(\nabla_Z(\nabla_{\varphi X} f)) - (df)(R(Z, Y)\varphi X) \\
&= (\nabla^3 f)(\varphi X, Z, Y) - (df)(R(Z, Y)\varphi X)
\end{aligned} \tag{2.23}$$

Since h is symmetric and the curvature tensor R of g is pure with respect to φ, from (2.22) and (2.23), we have

$$(\nabla^3 f)(Z, \varphi X, Y) = (\nabla^3 f)(Z, X, \varphi Y), \tag{2.24}$$

i.e. a tensor field $\nabla^3 f$ is pure in all arguments. On the other hand,

$$\big(\Phi_\varphi h\big)(X, Z_1, Z_2)$$

$$\begin{aligned}
&= (\varphi X)(h(Z_1, Z_2)) - X(h(\varphi Z_1, Z_2)) - h\big(\nabla_{\varphi X} Z_1, Z_2\big) \\
&\quad + h((\nabla\varphi)(X, Z_1), Z_2) + h(Z_1, (\nabla\varphi)(X, Z_2)) - h\big(Z_1, \nabla_{\varphi X} Z_2\big) \\
&\quad + h(\varphi(\nabla_X Z_1), Z_2) + h(\varphi Z_1, \nabla_X Z_2) \\
&= \big(\nabla_{\varphi X} h\big)(Z_1, Z_2) - (\nabla_X h)(\varphi Z_1, Z_2) + h((\nabla\varphi)(X, Z_1), Z_2) \\
&\quad + h(Z_1, (\nabla\varphi)(X, Z_2)) \qquad (2.25)
\end{aligned}$$

Substituting $h(Z_1, Z_2) = \nabla_{Z_1}\nabla_{Z_2} f$ and $\nabla\varphi = 0$ in (2.25), by virtue of (2.24), we have

$$\begin{aligned}
&\big(\Phi_\varphi h\big)(X, Z_1, Z_2) = \big(\Phi_\varphi h\big)(X, Z_2, Z_1) \\
&= \big(\nabla_{\varphi X}(\nabla^2 f)\big)(Z_2, Z_1) - \big(\nabla_X(\nabla^2 f)\big)(Z_2, \varphi Z_1), \\
&= \big(\nabla^3 f\big)(Z_2, Z_1, \varphi X) - \big(\nabla^3 f\big)(Z_2, \varphi Z_1, X) = 0
\end{aligned}$$

i.e. h is holomorphic. Then, using Theorem 2.1, we see that ${}^h\widetilde{\nabla}\varphi = 0$. Thus, we have

Theorem 2.11 *The Hessian-Norden triple* $(M_{2r}, h = \nabla^2 f, J)$ *is a Kähler-Norden manifold.*

2.4 Twin Norden Metric Connections

It is well known that the pair (J, g) of an almost Hermitian structure defines a fundamental 2-form Ω by $\Omega(X, Y) = g(JX, Y)$. Let ∇ be the Levi–Civita connection of g. If the skew-symmetric tensor Ω is a Killing-Yano tensor, i.e.

$$(\nabla_X \Omega)(Y, Z) + (\nabla_Y \Omega)(X, Z) = 0 \qquad (2.26)$$

or equivalently, if the almost complex structure J satisfies $(\nabla_X J)Y + (\nabla_Y J)X = 0$ for any $X, Y \in \mathfrak{I}_0^1(M_{2r})$, then the manifold is called a nearly Kähler manifold (also known as K-spaces or almost Tachibana spaces).

Let now (M, g, J) be an almost Norden manifold. Then the pair (J, g) defines, as usual, the twin Norden metric $G(Y, Z) = (g \circ J)(Y, Z) = g(JY, Z)$, but G is symmetric, rather than a 2-form Ω. Thus, the Norden pair (J, g) does not give rise to a 2-form, and the Killing-Yano Eq. (2.26) has no immediate meaning. Therefore, we can replace the Killing-Yano equation by *Codazzi equation*

$$(\nabla_X G)(Y, Z) - (\nabla_Y G)(X, Z) = 0. \qquad (2.27)$$

Equation (2.27) is equivalent to

$$(\nabla_X J)Y - (\nabla_Y J)X = 0. \qquad (2.28)$$

Theorem 2.12 *Let the triple (M, g, J) be an almost Norden manifold and G be a twin Norden metric which satisfies the Codazzi* Eq. (2.27). *Then J is integrable.*

Proof Using $\nabla_X Y - \nabla_Y X = [X, Y]$, (2.28) and

$$\begin{aligned}(\nabla_X J)(JY) &= \nabla_X(J(JY)) - J(\nabla_X JY)\\ &= -\nabla_X Y - J((\nabla_X J)Y + J(\nabla_X Y))\\ &= -\nabla_X Y - J(\nabla_X J)Y - J^2(\nabla_X Y) = -J(\nabla_X J)Y,\end{aligned} \tag{2.29}$$

we have

$$\begin{aligned}N_J(X, Y) &= [JX, JY] - J[X, JY] - J[JX, Y] - [X, Y]\\ &= \nabla_{JX} JY - \nabla_{JY} JX - J(\nabla_X JY - \nabla_{JY} X)\\ &\quad - J(\nabla_{JX} Y - \nabla_Y JX) + J^2(\nabla_X Y - \nabla_Y X)\\ &= -J((\nabla_X J)Y - (\nabla_Y J)X) + (\nabla_{JX} J)Y - (\nabla_{JY} J)X\\ &= (\nabla_Y J)JX - (\nabla_X J)JY = -J((\nabla_Y J)X - (\nabla_X J)Y) = 0,\end{aligned}$$

i.e. the Nijenhuis tensor N_J vanishes. The proof of Theorem 2.12 is complete.

In the above sections, we have given the Norden metric g and considered exclusively the Levi–Civita connection ∇ of g. This is the unique connection which satisfies $\nabla g = 0$ and has no torsion. But there are many other connections $\widetilde{\nabla}$ with torsion parallelizing the metric g. We call these connections *Norden metric connections with torsion.*

Let (M, g, J) be an almost Norden manifold. If we introduce a connection $\widetilde{\nabla}$ by

$$\widetilde{\nabla}_X Y = \nabla_X Y + S(X, Y)$$

for any $X, Y \in \Im_0^1(M_{2r})$, where S is a tensor field of type (1.2), then the torsion tensor T of $\widetilde{\nabla}$ is given by

$$\begin{aligned}T(X, Y) &= \widetilde{\nabla}_X Y - \widetilde{\nabla}_Y X - [X, Y]\\ &= \nabla_X Y + S(X, Y) - \nabla_Y X - S(Y, X) - [X, Y]\\ &= T^{\nabla}(X, Y) + S(X, Y) - S(Y, X) = S(X, Y) - S(Y, X)(T^{\nabla} = 0).\end{aligned} \tag{2.30}$$

For the covariant derivative $\widetilde{\nabla}$ of g, we have

$$\begin{aligned}(\widetilde{\nabla}_X g)(Y, Z) &= X(g(Y, Z)) - g\big(\widetilde{\nabla}_X Y, Z\big) - g\big(Y, \widetilde{\nabla}_X Z\big)\\ &= X(g(Y, Z)) - g(\nabla_X Y + S(X, Y), Z) - g(Y, \nabla_X Z + S(X, Z))\\ &= (\nabla_X g)(Y, Z) - g(S(X, Y), Z) - g(Y, S(X, Z))\\ &= -g(S(X, Y), Z) - g(Y, S(X, Z)).\end{aligned}$$

Consequently, in order to have $\widetilde{\nabla} g = 0$, it is necessary and sufficient that

$$g(S(X, Y), Z) + g(Y, S(X, Z)) = 0 .$$

From here, we have

Theorem 2.13 *Let (M, g, J) be an almost Norden manifold. A connection $\widetilde{\nabla} = \nabla + S$ is a Norden metric connection with torsion T (i.e. $\widetilde{\nabla} g = 0$) if and only if*

$$S(X, Y, Z) + S(X, Z, Y) = 0 , \tag{2.31}$$

where $S(X, Y, Z) = g(S(X, Y), Z)$.

Now putting $T(X, Y, Z) = g(T(X, Y), Z)$, from (2.30), we have

$$T(X, Y, Z) = S(X, Y, Z) - S(Y, X, Z) .$$

Similarly,

$$T(Z, X, Y) = S(Z, X, Y) - S(X, Z, Y) ,$$

$$T(Z, Y, X) = S(Z, Y, X) - S(Y, Z, X) .$$

Using (2.31), from the last equations, we obtain

$$S(X, Y, Z) = 1/2(T(X, Y, Z) + T(Z, X, Y) + T(Z, Y, X)) .$$

Let now $G(Y, Z) = (g \circ J)(Y, Z) = g(JY, Z)$ be a twin Norden metric. Then, in order that to have $\widetilde{\nabla} G = 0$, it is necessary and sufficient that we have

$$\begin{aligned}
\big(\widetilde{\nabla}_X G\big)(Y, Z) &= (\nabla_X G)(Y, Z) - G(S(X, Y), Z) - G(Y, S(X, Z)) \\
&= (\nabla_X G)(Y, Z) - g(JS(X, Y), Z) - g(JY, S(X, Z)) \\
&= (\nabla_X G)(Y, Z) - g(S(X, Y), JZ) - g(S(X, Z), JY) \\
&= (\nabla_X G)(Y, Z) - S(X, Y, JZ) - S(X, Z, JY) = 0
\end{aligned}$$

which is equivalent to

$$(\nabla_X G)(Y, Z) - S'_J(X, Y, Z) - S'_J(X, Z, Y) = 0, \tag{2.32}$$

where $S'_J(X, Y, Z) = S(X, Y, JZ)$.

The connection $\widetilde{\nabla}$ is not completely determined by (2.31) and (2.32). So we can introduce some other condition on S. We try to solve the equation with respect to S. From now on, we assume only $\widetilde{\nabla} G = 0$ and make no use of $\widetilde{\nabla} g = 0$ (i.e.

$S(X, Y, Z) + S(X, Z, Y) = 0$). This latter equation will be satisfied in special cases as a consequence of the equation introduced in next part of section.

Case 1. The connection $\widetilde{\nabla} = \nabla + S$ is called a *twin Norden metric connection of type I* if

$$\widetilde{\nabla} G = 0$$

and

$$S'_J(X, Y, Z) - S'_J(X, Z, Y) = 0\,. \tag{2.33}$$

From (2.32) and (2.33), we have

$$S'_J(X, Y, Z) = {}^1\!/\!_2(\nabla_X G)(Y, Z)$$

from which

$$\begin{aligned} S(X, Y, JZ) &= {}^1\!/\!_2 g((\nabla_X J)Y, Z), \\ g(S(X, Y), JZ) &= {}^1\!/\!_2 g((\nabla_X J)Y, Z)\,, \\ g(JS(X, Y), Z) &= {}^1\!/\!_2 g((\nabla_X J)Y, Z), \\ JS(X, Y) &= {}^1\!/\!_2(\nabla_X J)Y \end{aligned} \tag{2.34}$$

or

$$S(X, Y) = -{}^1\!/\!_2 J(\nabla_X J)Y. \tag{2.35}$$

If we substitute JZ into Z in the second equation of (2.34), we have

$$S(X, Y, Z) = -{}^1\!/\!_2 g((\nabla_X J)Y, JZ) \tag{2.36}$$

On the other hand, using

$$\begin{aligned} &g((\nabla_X J)Z, JY) - g(Z, (\nabla_X J)JY) = g(\nabla_X JZ - J\nabla_X Z, JY) \\ &- g(Z, -\nabla_X Y - J(\nabla_X JY)) = g(\nabla_X JZ, JY) + g(\nabla_X Z, Y) \\ &+ g(Z, \nabla_X Y) + g(JZ, \nabla_X JY) = Xg(Z, Y) - (\nabla_X g)(Z, Y) \\ &+ Xg(JZ, JY) - (\nabla_X g)(JZ, JY) = Xg(Z, Y) - Xg(Z, Y) = 0 \end{aligned}$$

and (2.29), we have

$$g((\nabla_X J)Z, JY) = g(Z, (\nabla_X J)JY) = -g(Z, J((\nabla_X J)Y)) = -g((\nabla_X J)Y, JZ),$$

which follows

$$S(X, Z, Y) + S(X, Y, Z) = -{}^1\!/\!_2(g((\nabla_X J)Y, JZ) + g((\nabla_X J)Z, JY)) = 0.$$

Thus, in an almost Norden manifold, the tensor S in the form (2.36) satisfies Eq. (2.31) (i.e. $\widetilde{\nabla} g = 0$), and consequently, the connection $\widetilde{\nabla} = \nabla - {}^1\!/_2 J(\nabla J)$ is a Norden metric connection with torsion. Thus, we have.

Theorem 2.14 *Every almost Norden manifold (M, g, J) admits a twin Norden metric connection of type I in the form $\widetilde{\nabla} = \nabla - {}^1\!/_2 J(\nabla J)$, and such connection is also a Norden metric connection with torsion.*

From (2.30) and (2.35), we see that the torsion tensor of the connection $\widetilde{\nabla} = \nabla + S$ is given by

$$T(X, Y) = -{}^1\!/_2 J((\nabla_X J)Y - (\nabla_Y J)X). \tag{2.37}$$

Let now the triple (M, g, J) be an almost Norden manifold with the Codazzi Eq. (2.27). Then from (2.28) and (2.37), we find that $T = 0$, i.e. the twin Norden metric connection of type I reduces to Levi–Civita connection. Thus, we have [55] (see also [17]).

Theorem 2.15 *If an almost Norden manifold satisfies the Codazzi* Eq. (2.27), *then the twin Norden metric connection of type I coincides with the Levi–Civita connection of g, i.e. the metric g and the twin metric G share the same Levi–Civita connection.*

Case 2. The connection $\widetilde{\nabla} = \nabla + S$ is called a *twin Norden metric connection of type II* if

$$\widetilde{\nabla} G = 0$$

and

$$S'_J(X, Y, Z) - S'_J(Z, Y, X) = 0\,. \tag{2.38}$$

From (2.32), we have

$$\begin{aligned}
(\nabla_X G)(Y, Z) - S'_J(X, Y, Z) - S'_J(X, Z, Y) = 0\,,\\
(\nabla_Y G)(Z, X) - S'_J(Y, Z, X) - S'_J(Y, X, Z) = 0\,,\\
(\nabla_Z G)(X, Y) - S'_J(Z, X, Y) - S'_J(Z, Y, X) = 0\,,
\end{aligned}$$

and consequently, taking account of (2.38), we find

$$\begin{aligned}
&(\nabla_X G)(Y, Z) - (\nabla_Y G)(Z, X) + (\nabla_Z G)(X, Y) = 2S'_J(X, Y, Z)\\
&= 2S(X, Y, JZ) = 2g(S(X, Y), JZ) = 2g(JS(X, Y), Z).
\end{aligned} \tag{2.39}$$

Since the operator Φ_J applied to g reduces to the following form (see (2.8)):

$$(\Phi_J g)(Y, Z, X) = (\nabla_X G)(Y, Z) - (\nabla_Y G)(Z, X) + (\nabla_Z G)(X, Y)$$

and the Kähler-Norden condition ($\nabla J = 0$) is equivalent to $\Phi_J g = 0$ (see Theorem 2.1), from (2.39), we have $S = 0$, the twin Norden metric connection $\widetilde{\nabla}$ reduces to Levi–Civita connection ∇, it is clear that the tensor $S = 0$ satisfies Eq. (2.31), and consequently, the connection $\widetilde{\nabla} = \nabla$ is a torsion-free Norden metric connection. Thus, we have

Theorem 2.16 *If an almost Norden manifold (M, g, J) is Kähler-Norden, then the twin Norden metric connection $\widetilde{\nabla}$ of type II coincides with the torsion-free Norden metric connection, i.e. with the Levi–Civita connection ∇.*

Let now

$$(\nabla_Y G)(Z, X) = (\nabla_Z G)(Y, X).$$

Then from (2.29) and (2.39), we find

$$\begin{aligned} 2g(JS(X, Y), Z) &= (\nabla_X G)(Y, Z) - (\nabla_Y G)(Z, X) + (\nabla_Z G)(Y, X) \\ &= (\nabla_X G)(Y, Z) = g((\nabla_X J)Y, Z) \end{aligned}$$

or

$$S(X, Y) = -{}^1\!/\!_2 J(\nabla_X J)Y\,. \tag{2.40}$$

By similar devices as above (Case 1, Theorem 2.14), we easily see that the tensor S in the form (2.40) satisfies Eq. (2.31), and consequently, the twin Norden metric connection $\widetilde{\nabla}$ of type II is given by $\widetilde{\nabla} = \nabla - {}^1\!/\!_2 J(\nabla J)$, i.e. coincides with the type of I and also it reduces to the Levi–Civita connection $\widetilde{\nabla} = \nabla$. Thus, we have.

Theorem 2.17 *If an almost Norden manifold satisfies the Codazzi* Eq. (2.27), *then the twin Norden metric connection of type II coincides with the Levi–Civita connection ∇ of g.*

Let now G be a Killing symmetric tensor, i.e. $\underset{X,Y,Z}{\sigma} (\nabla_X G)(Y, Z) = 0$, where σ is the cyclic sum with respect to X, Y and Z. This is the class of the quasi-Kähler manifold [32, 38]. The structure J on such manifolds is nonintegrable.

Similarly, if (M, g, J) is quasi-Kähler, i.e.

$$(\nabla_X G)(Y, Z) + (\nabla_Y G)(Z, X) + (\nabla_Z G)(X, Y) = 0,$$

then from (2.39), we find

$$0 = (\nabla_X G)(Y, Z) + (\nabla_Y G)(Z, X) + (\nabla_Z G)(X, Y)$$

$$
\begin{aligned}
&= (\nabla_X G)(Y, Z) - (\nabla_Y G)(Z, X) + (\nabla_Z G)(X, Y) + 2(\nabla_Y G)(Z, X) \\
&= 2g(JS(X, Y), Z) + 2(\nabla_Y G)(Z, X) = 2g(JS(X, Y), Z) + 2(\nabla_Y G)(X, Z) \\
&= 2g(JS(X, Y), Z) + 2g((\nabla_Y J)X, Z),
\end{aligned}
$$

i.e.

$$g(JS(X, Y), Z) = -g((\nabla_Y J)X, Z) \Rightarrow S(X, Y) = J(\nabla_Y J)X.$$

It is clear that the tensor S satisfies (2.31) (see Case 1, Theorem 2.14). Thus, we have

Theorem 2.18 *Every quasi-Kähler manifold (M, g, J) admits a twin Norden metric connection of type II of the form $\widetilde{\nabla} = \nabla + J(\nabla J)$, and such connection is also a Norden metric connection with torsion.*

Remark 2.2. Given an almost Hermitian manifold (M, g, J), there is a unique connection $\widetilde{\nabla}$ (known as the Bismut connection) with totally skew torsion which preserves both the complex structure and the Hermitian metric, i.e. $\widetilde{\nabla} g = 0$ and $\widetilde{\nabla} J = 0$. For the almost Norden manifolds, from $\widetilde{\nabla} g = 0$ and $\widetilde{\nabla} G = 0$, we have $\widetilde{\nabla} J = 0$; therefore, in some aspects, Norden metric connections of types I and II introduced in the present section are similar to Bismut connection.

2.5 Norden-Walker Manifolds

The main purpose of the present section is to study the Norden metrics on 4-dimensional Walker manifolds.

A neutral metric g on a 4-manifold M_4 is said to be a *Walker metric* if there exists a 2-dimensional null distribution D on M_4, which is parallel with respect to g. From Walker's theorem [84], there is a system of coordinates (x, y, z, t) with respect to which g takes the following local canonical form

$$g = (g_{ij}) = \begin{pmatrix} 0 & 0 & 1 & 0 \\ 0 & 0 & 0 & 1 \\ 1 & 0 & a & c \\ 0 & 1 & c & b \end{pmatrix}, \tag{2.41}$$

where a, b, c are smooth functions of the coordinates (x, y, z, t). The parallel null 2-plane D is spanned locally by $\{\partial_x, \partial_y\}$, where $\partial_x = \frac{\partial}{\partial x}$, $\partial_y = \frac{\partial}{\partial y}$. Walker manifolds are in focus of many authors for investigations of different geometrical problems (see, for example, [2–4, 6, 11, 12, 33–37, 53, 54, 63, 64, 66]).

In [35], a proper almost complex structure with respect to g is defined as a g-orthogonal almost complex structure J so that J is a standard generator of a positive $\frac{\pi}{2}$ rotation on D, i.e. $J\partial_x = \partial_y$ and $J\partial_y = -\partial_x$. Then for the Walker metric g, such a proper almost complex structure J is determined uniquely as

$$J = \begin{pmatrix} 0 & -1 & -c & \frac{1}{2}(a-b) \\ 1 & 0 & \frac{1}{2}(a-b) & c \\ 0 & 0 & 0 & -1 \\ 0 & 0 & 1 & 0 \end{pmatrix} \tag{2.42}$$

In [6], for such a proper almost complex structure J on Walker 4-manifold M, an almost Norden structure (g^{N+}, J) is constructed, where g^{N+} is a metric on M, with properties $g^{N+}(JX, JY) = -g^{N+}(X, Y)$. In fact, as one of these examples, such a metric takes the form (see Proposition 6 in [6], unfortunately, the calculations of the component g^{N+}_{44} in [6] are erroneous):

$$g^{N+} = \begin{pmatrix} 0 & -2 & 0 & -b \\ -2 & 0 & -a & -2c \\ 0 & -a & 0 & \frac{1}{2}(1-ab) \\ -b & -2c & \frac{1}{2}(1-ab) & -2bc \end{pmatrix}.$$

We may call this an *almost Norden-Walker metric*. The construction of such a structure in [6] is to find a Norden metric for a given almost complex structure, which is different from the Walker metric.

The purpose of the present section is to find also an almost Norden-Walker structure (g, F), where the metric is nothing but the Walker metric g, with an appropriate almost complex structure F, to be determined. That is, for a fixed metric g, we will find an almost complex structure F which satisfies $g(FX, FY) = -g(X, Y)$. In [6], for a given almost complex structure, a metric is constructed. Our method is, however, for a given metric, an almost complex structure is constructed.

Let F be an almost complex structure on a Walker manifold M_4, which satisfies

(i) $F^2 = -I$,
(ii) $g(FX, Y) = g(X, FY)$
(iii) $F\partial_x = \partial_y, \; F\partial_y = -\partial_x$.

We easily see that these three properties define F nonuniquely, i.e.

$$\begin{cases} F\partial_x = \partial_y, \\ F\partial_y = -\partial_x, \\ F\partial_z = \alpha\partial_x + \frac{1}{2}(a+b)\partial_y - \partial_t, \\ F\partial_t = -\frac{1}{2}(a+b)\partial_x + \alpha\partial_y + \partial_z, \end{cases}$$

and F has the local components

$$F = (F_j^i) = \begin{pmatrix} 0 & -1 & \alpha & -\frac{1}{2}(a+b) \\ 1 & 0 & \frac{1}{2}(a+b) & \alpha \\ 0 & 0 & 0 & 1 \\ 0 & 0 & -1 & 0 \end{pmatrix}$$

with respect to the natural frame $\{\partial_x, \partial_y, \partial_z, \partial_t\}$, where $\alpha = \alpha(x, y, z, t)$ is an arbitrary function. We must note that the proper almost complex structure J as in (2.42) is determined uniquely. In our case of the almost Norden-Walker structure, the almost complex structure F just obtained contains an arbitrary function $\alpha(x, y, z, t)$. Our purpose is to find a nontrivial almost Norden-Walker structure with the Walker metric g explicitly. Therefore, we now put $\alpha = c$. Then g defines a unique almost complex structure

$$\varphi = (\varphi_j^i) = \begin{pmatrix} 0 & -1 & c & -\frac{1}{2}(a+b) \\ 1 & 0 & \frac{1}{2}(a+b) & c \\ 0 & 0 & 0 & 1 \\ 0 & 0 & -1 & 0 \end{pmatrix}. \tag{2.43}$$

The triple (M_4, φ, g) is called *almost Norden-Walker manifold*. In conformity with the terminology of [34, 35], we call φ the *proper almost complex structure.*

Remark 2.3 From (2.43), we immediately see that in the case $a = -b$ and $c = 0$, φ is integrable.

We now consider the general case for integrability. The almost complex structure φ on almost Norden-Walker manifolds is integrable if and only if

$$(N_\varphi)_{jk}^i = \varphi_j^m \partial_m \varphi_k^i - \varphi_k^m \partial_m \varphi_j^i - \varphi_m^i \partial_j \varphi_k^m + \varphi_m^i \partial_k \varphi_j^m = 0. \tag{2.44}$$

From (2.43) and (2.44), we find the following integrability condition:

Theorem 2.19 *The proper almost complex structure φ on almost Norden-Walker manifolds is integrable if and only if the following PDEs hold:*

$$\begin{cases} a_x + b_x + 2c_y = 0, \\ a_y + b_y - 2c_x = 0. \end{cases} \tag{2.45}$$

From this theorem, we see that if $a = -b$ and $c = 0$, then φ is integrable (see Remark 2.3).

Let (M_4, φ, g) be an integrable almost Norden-Walker manifold ($N_\varphi = 0$) and $a = b$. Then Eq. (2.45) reduces to

$$\begin{cases} a_x = -c_y, \\ a_y = c_x, \end{cases} \tag{2.46}$$

from which follows

$$a_{xx} + a_{yy} = 0,$$
$$c_{xx} + c_{yy} = 0,$$

i.e. the functions a and c are harmonic with respect to the arguments x and y. Thus, we have

Theorem 2.20 *If the triple (M_4, φ, g) is an integrable almost Norden-Walker manifold and $a = b$, then a and c are harmonic with respect to the arguments x, y.*

Example 2.1 We now apply the Theorem 2.20 to establish the existence of special types of Norden-Walker metrics. In our arguments, the harmonic function plays an important role.

Let $a = b$ and $h(x, y)$ be a harmonic function of variables x and y, for example, $h(x, y) = e^x \cos y$. We put

$$a = a(x, y, z, t) = h(x, y) + \alpha(z, t) = e^x \cos y + \alpha(z, t),$$

where α is an arbitrary smooth function of z and t. Then, a is also hormonic with respect to x and y. We have

$$a_x = e^x \cos y,$$
$$a_y = -e^x \sin y.$$

From (3.46), we have PDEs for c to satisfy as

$$c_x = a_y = -e^x \sin y,$$
$$c_y = -a_x = -e^x \cos y.$$

These PDEs have the solutions

$$c = -e^x \sin y + \beta(z, t),$$

where β is arbitrary smooth function of z and t. Thus, the Norden-Walker metric reduces to the following form:

$$g = (g_{ij}) = \begin{pmatrix} 0 & 0 & 1 & 0 \\ 0 & 0 & 0 & 1 \\ 1 & 0 & e^x \cos y + \alpha(z,t) & -e^x \sin y + \beta(z,t) \\ 0 & 1 & -e^x \sin y + \beta(z,t) & e^x \cos y + \alpha(z,t) \end{pmatrix}.$$

2.6 Kähler-Norden-Walker Manifolds

Let (M_4, φ, g) be an almost Norden-Walker manifold. If

$$\begin{aligned}(\Phi_\varphi g)_{kij} &= \varphi_k^m \partial_m g_{ij} - \varphi_i^m \partial_k g_{mj} \\ &\quad + g_{mj}\left(\partial_i \varphi_k^m - \partial_k \varphi_i^m\right) + g_{im} \partial_j \varphi_k^m = 0 \end{aligned} \tag{2.47}$$

then by virtue of Theorem 2.1, φ is integrable and the triple (M_4, φ, g) is called a holomorphic Norden-Walker or a Kähler-Norden-Walker manifold. Taking account of Theorem 2.2, we see that Kähler-Norden-Walker manifold with condition $\Phi_\varphi g = 0$ and $N_\varphi \neq 0$ does not exist.

Substituting (2.41) and (2.43) in (2.47), we see that the nonvanishing components of $\Phi_\varphi g$ are

$$\begin{aligned}
&(\Phi_\varphi g)_{xzz} = a_y, \\
&(\Phi_\varphi g)_{xzt} = (\Phi_\varphi g)_{xzz} = \frac{1}{2}(b_x - a_x) + c_y, \\
&(\Phi_\varphi g)_{xtt} = b_y - 2c_x, \\
&(\Phi_\varphi g)_{yzz} = -a_x, \\
&(\Phi_\varphi g)_{yzt} = (\Phi_\varphi g)_{ytz} = \tfrac{1}{2}(b_y - a_y) - c_x, \\
&(\Phi_\varphi g)_{ytt} = -b_x - 2c_y, \\
&(\Phi_\varphi g)_{zxx} = (\Phi_\varphi g)_{zxx} = (\Phi_\varphi g)_{txt} = (\Phi_\varphi g)_{ttx} = c_x, \\
&(\Phi_\varphi g)_{zxt} = (\Phi_\varphi g)_{ztx} = -(\Phi_\varphi g)_{txz} = -(\Phi_\varphi g)_{tzx} = \tfrac{1}{2}(a_x + b_x), \\
&(\Phi_\varphi g)_{zyz} = (\Phi_\varphi g)_{zzy} = (\Phi_\varphi g)_{tyt} = (\Phi_\varphi g)_{tty} = c_y, \\
&(\Phi_\varphi g)_{z,t} = (\Phi_\varphi g)_{zty} = -(\Phi_\varphi g)_{tzz} = -(\Phi_\varphi g)_{tyy} = \frac{1}{2}(a_y + b_y), \\
&(\Phi_\varphi g)_{zzz} = ca_x - a_t + 2c_z + \frac{1}{2}(a+b)a_y, \\
&(\Phi_\varphi g)_{zzt} = (\Phi_\varphi g)_{ztz} = cc_x + b_z + \frac{1}{2}(a+b)c_y,
\end{aligned}$$

$$\begin{aligned}
(\Phi_\varphi g)_{ztt} &= cb_x + a_t - 2c_z + \tfrac{1}{2}(a+b)b_y,\\
(\Phi_\varphi g)_{tzz} &= ca_y - b_z - \frac{1}{2}(a+b)a_x,\\
(\Phi_\varphi g)_{tzt} &= (\Phi_\varphi g)_{ttz} = cc_y - a_t + 2c_z - \tfrac{1}{2}(a+b)c_x,\\
(\Phi_\varphi g)_{ttt} &= cb_y + b_z - \tfrac{1}{2}(a+b)b_x.
\end{aligned} \tag{2.48}$$

From (2.48), we have.

Theorem 2.21 *The triple (M_4, φ, g) is Kähler-Norden-Walker if and only if the following PDEs hold:*

$$a_x = a_y = c_x = c_y = b_x = b_y = b_z = 0, \qquad a_t - 2c_z = 0. \tag{2.49}$$

Remark 2.4 The triple (M_4, φ, g) with metric

$$g = (g_{ij}) = \begin{pmatrix} 0\,0\,1 & 0 \\ 0\,0\,0 & 1 \\ 1\,0\,a(z) & 0 \\ 0\,1\,0 & b(t) \end{pmatrix}$$

is always Kähler-Norden-Walker.

2.7 On Curvatures of Norden-Walker Metrics

If R and r are, respectively, the curvature tensor and the scalar curvature of the Walker metric, then the nonvanishing components of R and r have, respectively, expressions (see [34, 35])

$$\begin{aligned}
&R_{xxzz} = -\frac{1}{2}a_{xx}, \quad R_{xxt} = -\frac{1}{2}c_{xx}, \quad R_{xyzz} = -\frac{1}{2}a_{xy}, \quad R_{xyyt} = -\frac{1}{2}c_{xy},\\
&R_{xzzt} = \frac{1}{2}a_{xt} - \frac{1}{2}c_{xz} - \frac{1}{4}a_y b_x + \frac{1}{4}c_x c_y, \quad R_{xxtt} = -\frac{1}{2}b_{xx}, \quad R_{xxyz} = -\frac{1}{2}c_{xy},\\
&R_{xxty} = -\frac{1}{2}b_{xy}, \quad R_{xttt} = \frac{1}{2}c_{xt} - \frac{1}{2}b_{xz} - \frac{1}{4}(c_x)^2 + \frac{1}{4}a_x b_x - \frac{1}{4}b_x c_y + \frac{1}{4}b_y c_x,\\
&R_{yzyz} = -\frac{1}{2}a_{yy}, \quad R_{yyyt} = -\frac{1}{2}c_{yy}\\
&R_{yzzt} = \frac{1}{2}a_{yt} - \frac{1}{2}c_{yz} - \frac{1}{4}a_x c_y + \frac{1}{4}a_y c_x - \frac{1}{4}a_y b_y + \frac{1}{4}(c_y)^2, \quad R_{ybtt} = -\frac{1}{2}b_{yy},\\
&R_{ytzt} = \tfrac{1}{2}c_{yt} - \tfrac{1}{2}b_{yz} - \tfrac{1}{4}c_x c_y + \tfrac{1}{4}a_y b_x\\
&R_{ztzt} = c_{zt} - \tfrac{1}{2}a_{tt} - \tfrac{1}{2}b_{zz} - \tfrac{1}{4}a(c_x)^2 + \tfrac{1}{4}aa_x b_x + \tfrac{1}{4}ca_x b_y - \tfrac{1}{2}cc_x c_y - \tfrac{1}{2}a_t c_x
\end{aligned}$$

$$+ \tfrac{1}{2}a_x c_t - \tfrac{1}{4}a_x b_z + \tfrac{1}{4}ca_y b_x + \tfrac{1}{4}ba_y b_y - \tfrac{1}{4}b(c_y)^2 - \tfrac{1}{2}b_z c_y$$
$$+ \tfrac{1}{4}a_y b_t + \tfrac{1}{4}a_z b_x + \tfrac{1}{2}b_y c_z - \tfrac{1}{4}a_t b_y \tag{2.50}$$

and

$$r = a_{xx} + 2c_{xy} + b_{yy}. \tag{2.51}$$

Suppose that the triple (M_4, φ, g) is Kähler-Norden-Walker. Then from (2.49) and (2.50), we see that

$$R_{ztzt} = c_{zt} - \tfrac{1}{2}a_{tt} = -\tfrac{1}{2}(a_t - 2c_z)_t = 0.$$

and the other components of R directly all vanish. Thus, we have

Theorem 2.22 *If a Norden-Walker manifold (M_4, φ, g) is Kähler-Norden-Walker, then M_4 is flat.*

Let (M_4, φ, g) be a Norden-Walker manifold with the integrable proper structure φ, i.e. $N_\varphi = 0$. If $a = b$, then from the proof of Theorem 2.19, we see that Eq. (2.46) holds. If $c = c(y, z, t)$ and $c = c(x, z, t)$, then $c_{xy} = (c_x)_y = (c_y)_x = 0$, and by virtue of (2.46), we find $a = a(x, z, t)$ and $a = (y, z, t)$, respectively. Using of $c_{xy} = 0$ and $a_{xx} + a_{yy} = 0$, we from (2.51) obtain $r = 0$. Thus, we have the following result:

Theorem 2.23 *If (M_4, φ, g) and $(M_4, \varphi, \tilde{g})$ are a Norden-Walker manifolds with the integrable proper structure φ and with metrics*

$$g = \begin{pmatrix} 0\,0\,1 & 0 \\ 0\,0\,0 & 1 \\ 1\,0\;a(x,z,t) & c(y,z,t) \\ 0\,1\;c(y,z,t) & a(x,z,t) \end{pmatrix}, \quad \tilde{g} = \begin{pmatrix} 0\,0\,1 & 0 \\ 0\,0\,0 & 1 \\ 1\,0\;a(y,z,t) & c(x,z,t) \\ 0\,1\;c(x,z,t) & a(y,z,t) \end{pmatrix},$$

then both (M_4, φ, g) and $(M_4, \varphi, \tilde{g})$ are scalar flat.

2.8 Isotropic Anti-Kähler Manifolds

It is well known that the inner product in the vector space can be extended to an inner product in the tensor space. In fact, if $\underset{1}{t}$ and $\underset{2}{t}$ are tensors of type (r, s) with components $\underset{1}{t}{}^{i_1 \ldots i_r}_{j_1 \ldots j_s}$ and $\underset{2}{t}{}^{k_1 \ldots k_r}_{l_1 \ldots l_s}$, then.

$$g(\underset{1}{t},\underset{2}{t}) = g_{i_1k_1}\cdots g_{i_rk_r}g^{j_1l_1}\cdots g^{j_sl_s}\underset{1}{t}{}^{i_1\dots i_r}_{j_1\dots j_s}\underset{2}{t}{}^{k_1\dots k_r}_{l_1\dots l_s}$$

If $\underset{1}{t} = \underset{2}{t} = \nabla\varphi \in \Im^1_2(M_{2n})$, then the square norm $\|\nabla\varphi\|^2$ of $\nabla\varphi$ is defined by

$$\|\nabla\varphi\|^2 = g^{ij}g^{kl}g_{ms}(\nabla\varphi)^m_{ik}(\nabla\varphi)^s_{jl},$$

where ∇ is the Levi–Civita connection of g. An almost Norden manifold (M_4, φ, g) is said to be *isotropic anti-Kähler* if $\|\nabla\varphi\|^2 = 0$ (see [21]). It is clear that if the triple (M_4, φ, g) is Kähler-Norden, then it is isotropic anti-Kähler. Our purpose in this section is to show that an almost Norden-Walker manifold (M_4, φ, g) is isotropic anti-Kähler.

The inverse of the metric tensor (2.41) is given by

$$g^{-1} = (g^{ij}) = \begin{pmatrix} -a & -c & 1 & 0 \\ -c & -b & 0 & 1 \\ 1 & 0 & 0 & 0 \\ 0 & 1 & 0 & 0 \end{pmatrix}. \tag{2.52}$$

After some calculations, we see that the nonvanishing components of $\nabla\varphi$ are

$$\begin{aligned}
\nabla_x\varphi^x_z &= \nabla_x\varphi^y_t = c_x,\ \nabla_y\varphi^x_z = \nabla_y\varphi^y_t = c_y,\\
\nabla_z\varphi^x_x &= -\nabla_z\varphi^y_y = \nabla_z\varphi^z_z = -\nabla_z\varphi^t_t = \tfrac{1}{2}a_y + \tfrac{1}{2}c_x,\\
\nabla_z\varphi^y_x &= \nabla_z\varphi^x_y = \nabla_z\varphi^t_z = \nabla_z\varphi^z_t = -\tfrac{1}{2}a_x + \tfrac{1}{2}c_y,\\
\nabla_z\varphi^x_z &= 2c_z + ca_x - a_t - \tfrac{1}{2}cc_y - \tfrac{1}{2}ac_x + \tfrac{1}{2}ba_y,\\
\nabla_z\varphi^y_z &= a_z + \tfrac{1}{4}ac_y - \tfrac{1}{4}bc_y + ca_y + \tfrac{3}{4}aa_x + \tfrac{1}{4}ba_x,\\
\nabla_z\varphi^x_t &= \tfrac{1}{4}aa_x - \tfrac{1}{4}ba_x + ca_y + \tfrac{3}{4}bc_y + cc_x + \tfrac{1}{4}ac_y,\\
\nabla_z\varphi^y_t &= 2c_z + \tfrac{1}{2}cc_y - a_t + \tfrac{1}{2}ba_y + \tfrac{1}{2}ca_x - \tfrac{1}{2}ac_x,\\
\nabla_t\varphi^x_x &= -\nabla_t\varphi^y_y = \nabla_t\varphi^z_z = -\nabla_t\varphi^t_t = \tfrac{1}{2}c_y + \tfrac{1}{2}b_x,\\
\nabla_t\varphi^y_x &= \nabla_t\varphi^x_y = \nabla_t\varphi^t_z = \nabla_t\varphi^z_t = -\tfrac{1}{2}c_x + \tfrac{1}{2}b_y,\\
\nabla_t\varphi^x_z &= \tfrac{3}{2}cc_x + b_z - \tfrac{1}{2}cb_y - \tfrac{1}{2}ab_x + \tfrac{1}{2}bc_y,\\
\nabla_t\varphi^y_z &= \tfrac{1}{4}ab_y - \tfrac{1}{4}bb_y - \tfrac{1}{4}ac_x + \tfrac{1}{4}bc_x,\\
\nabla_t\varphi^x_t &= \tfrac{1}{4}ac_x - \tfrac{1}{4}bc_x + cc_y + \tfrac{1}{4}bb_y + cb_x - \tfrac{1}{4}ab_y,\\
\nabla_t\varphi^y_t &= \tfrac{1}{2}cb_y + b_z + \tfrac{1}{2}bc_y + \tfrac{1}{2}cc_x - \tfrac{1}{2}ab_x
\end{aligned} \tag{2.53}$$

Using (2.41), (2.52) and (2.53), we find

$$\|\nabla\varphi\|^2 = g^{ij}g^{kl}g_{ms}(\nabla\varphi)^m_{ik}(\nabla\varphi)^s_{jl} = 0.$$

Thus, we have

Theorem 2.24 *An almost Norden-Walker manifold (M_4, φ, g) is an isotropic anti-Kähler manifold.*

2.9 Quasi-Kähler-Norden-Walker Manifolds

The basis class of nonintegrable almost complex manifolds with Norden metric is the class of the quasi-Kähler manifolds. An almost Norden manifold (M_{2n}, φ, g) is called a *quasi-Kähler*, if

$$\underset{X,Y,Z}{\sigma} g((\nabla_X \varphi)Y, Z) = 0,$$

where σ is the cyclic sum by three arguments.

If we add $(\Phi_\varphi g)(X, Y, Z)$ and $(\Phi_\varphi g)(Z, Y, X)$ (see (2.8) or (2.9)), then by virtue of

$$g(Z, (\nabla_Y \varphi)X) = g((\nabla_Y \varphi)Z, X) \tag{2.54}$$

we find

$$(\Phi_\varphi g)(X, Y, Z) + (\Phi_\varphi g)(Z, Y, X) = 2g((\nabla_Y \varphi)Z, X) \tag{2.55}$$

Since

$$\begin{aligned}
&(\Phi_\varphi g)(X, Y, Z) = (\Phi_\varphi g)(X, Z, Y),\\
&(\Phi_\varphi g)(Y, Z, X) = -g((\nabla_Y \varphi)Z, X) + g((\nabla_Z \varphi)Y, X) + g(Z, (\nabla_X \varphi)Y),
\end{aligned}$$

from (2.54) and (2.55), we have

$$(\Phi_\varphi g)(X, Y, Z) + (\Phi_\varphi g)(Y, Z, X) + (\Phi_\varphi g)(Z, X, Y) = \underset{X,Y,Z}{\sigma} g((\nabla_X \varphi)Y, Z).$$

Thus, we have.

Theorem 2.25 *Let (M_{2n}, φ, g) be an almost Norden manifold. Then the Norden metric g is a quasi-Kähler-Norden if and only if*

$$(\Phi_\varphi g)(X, Y, Z) + (\Phi_\varphi g)(Y, Z, X) + (\Phi_\varphi g)(Z, X, Y) = 0, \tag{2.56}$$

for any $X, Y, Z \in \Im_0^1(M_{2n})$.

Equation (2.56) also can be written in the form:

$$(\Phi_\varphi g)(X, Y, Z) + 2g((\nabla_X \varphi)Y, Z) = 0.$$

From here we see that, if we take a local coordinate system, then a Norden-Walker manifold (M_4, φ, g) satisfying the condition of vanishing for $\Phi_k g_{ij} + 2\nabla_k G_{ij}$ is called a quasi-Kähler manifold, where G is defined by $G_{ij} = \varphi_i^m g_{mj}$. After some calculations, we see that the nonvanishing components of ∇G are

$$\begin{aligned}
&\nabla_x G_{zz} = \nabla_x G_{tt} = c_x, \nabla_y G_{zz} = \nabla_y G_{tt} = c_y,\\
&\nabla_z G_{xz} = \nabla_z G_{zx} = -\nabla_z G_{yt} = -\nabla_z G_{ty} = \frac{1}{2}a_y + \frac{1}{2}c_x,\\
&\nabla_z G_{xt} = \nabla_z G_{tx} = \nabla_z G_{yz} = \nabla_z G_{zy} = \frac{1}{2}c_y - \frac{1}{2}a_x,\\
&\nabla_z G_{zz} = 2c_z - a_t + \frac{1}{2}a_y(a+b) + ca_x,\\
&\nabla_z G_{zt} = \nabla_z G_{tz} = \frac{1}{2}ca_y + \frac{1}{2}cc_x - \frac{1}{4}a_x(a+b) + \frac{1}{4}c_y(a+b),\\
&\nabla_z G_{tt} = 2c_z - a_t + cc_y - \frac{1}{2}c_x(a+b),\\
&\nabla_t G_{xz} = \nabla_t G_{zx} = -\nabla_t G_{yt} = -\nabla_t G_{ty} = \frac{1}{2}b_x + \frac{1}{2}c_y,\\
&\nabla_t G_{xt} = \nabla_t G_{tx} = \nabla_t G_{yz} = \nabla_t G_{zy} = \frac{1}{2}b_y - \frac{1}{2}c_x,\\
&\nabla_t G_{zz} = b_z + cc_x + \frac{1}{2}c_y(a+b),\\
&\nabla_t G_{zt} = \nabla_t G_t = \frac{1}{2}cc_y + \frac{1}{2}cb_x - \frac{1}{4}c_x(a+b) + \frac{1}{4}b_y(a+b),\\
&\nabla_t G_{tt} = b_z + cb_y - \tfrac{1}{2}b_x(a+b).
\end{aligned} \tag{2.57}$$

From (2.48) and (2.57), we have.

Theorem 2.26 *A triple (M_4, φ, g) is a quasi-Kähler-Norden-Walker manifold if and only if the following PDEs hold:*

$$b_x = b_y = b_z = 0, a_y - 2c_x = 0, a_x - 2c_y = 0, ca_x - a_t + 2c_z - (a+b)c_x = 0.$$

2.10 The Goldberg Conjecture

Let now (M_4, φ, g) be an almost Hermitian manifold. The Goldberg conjecture (see [36]). states that an almost Hermitian manifold (M_4, φ, g) must be Kähler (or φ must be integrable) if the following three conditions are imposed: (G_1) if M is compact, (G_2) g is Einstein, and (G_3) if the fundamental 2-form $\Omega = g \circ \varphi$ is closed.

It should be noted that no progress has been made on the Goldberg conjecture, and the original conjecture is still an open problem. Despite many papers by various authors concerning the Goldberg conjecture, there is only Sekigawa paper [72] which obtained substantial results to the original Goldberg conjecture: let (M_4, φ, g) be an almost Hermitian manifold, which satisfies the three conditions (G_1),(G_2) and (G_3). If the scalar curvature of M is nonnegative, then φ must be integrable.

Let now $(M_4, \varphi, {}^w g)$ be an indefinite almost Kähler-Walker-Einstein compact manifold with the proper almost complex structure (2.42). As noted before, many examples of Norden-Walker metrics can be obtained by $g^{N+}(JX, JY) = -g^{N+}(X, Y)$ (see Sect. 2.5), and as one of these examples, such a metric has components

$$g^{N+} = \begin{pmatrix} 0 & -2 & 0 & -b \\ -2 & 0 & -a & -2c \\ 0 & -a & 0 & \frac{1}{2}(1-ab) \\ -b & -2c & \frac{1}{2}(1-ab) & -2bc \end{pmatrix}.$$

with respect to the Walker coordinates. Using Theorem 2.2, we have

Theorem 2.27 *The proper almost complex structure φ on indefinite almost Kähler-Walker-Einstein compact manifold $(M_4, \varphi, {}^w g)$ is integrable if $\Phi_\varphi g^{N+} = 0$, where g^{N+} is the induced Norden-Walker metric on M_4.*

This resolves the conjecture of Goldberg under the additional restriction on Norden-Walker metric ($g^{N+} \in Ker\ \Phi_\varphi$).

3 Problems of Lifts

In the first part of this chapter we focus on lifts from a manifold to its tensor bundle. Some introductory material concerning the tensor bundle is provided in Sect. 3.1. Section 3.2 is devoted to the study of the complete lifts of (1,1)-tensor fields along cross-setions in the tensor bundle. In Sect. 3.3 we study holomorphic cross-sections of tensor bundles.

In the second part we concentrate our attention to lifts from a manifold to its tangent bundles of order 1 and 2 by using the realization of holomorphic manifolds. The main purpose of Sects. 3.4–3.9 is to study the differential geometrical objects on the tangent bundle of order 1 corresponding to dual-holomorphic objects of the dual-holomorphic manifold. As a result of this approach, we find a new class of lifts, i.e. deformed complete lifts of functions, vector fields, forms, tensor fields and linear connections in the tangent bundle of order 1. Section 3.10 is devoted to the study of holomorphic metrics in the tangent bundle of order 2 (i.e. in the bundle of 2-jets) by using the Tachibana operator. By using the algebraic approach, the problem of deformed lifts of functions, vector fields and 1-forms is solved in Sects. 3.11 and 3.12. In Sect. 3.13, we investigate the complete lift of the almost complex structure to cotangent bundle and prove that it is a transfer by a symplectic isomorphism of complete lift to tangent bundle if the symplectic manifold with almost complex structure is an almost holomorphic A-manifold. Finally, in Sect. 3.14 we transfer via the differential of the musical isomorphism defined by pseudo-Riemannian metrics the complete lifts of vector fields and almost complex structures from the tangent bundle to the cotangent bundle.

Throughout the chapter we suppose that all tensor fields on manifolds and their bundles are of class C^∞.

A. Salimov, *Applications of Holomorphic Functions in Geometry*,
Frontiers in Mathematics, https://doi.org/10.1007/978-981-99-1296-4_3

3.1 Tensor Bundles

Let M_n be a differentiable manifold of class C^∞. Then the set $T_q^p(M_n) = \bigcup_{P \in M_n} T_q^p(P)$ is, by definition, the tensor bundle of type (p, q) over M_n, where $\cup$ denotes the disjoint union of the tensor spaces $T_q^p(P)$ for all $P \in M_n$. For any point $\tilde{P}$ of $T_q^p(M_n)$ the surjective correspondence $\tilde{P} \to P$ determines the natural projection $\pi : T_q^p(M_n) \to M_n$. In order to introduce a manifold structure in $T_q^p(M_n)$, we define local charts on it as follows: If x^j, $j = 1, \ldots, n$ are the local coordinates in a neighborhood U of $P \in M_n$, then a tensor t at P, i.e. a point $\tilde{P} = (P, t)$ which is an element of $T_q^p(M_n)$ is expressible in the form $(x^j, t_{j_1 \ldots j_q}^{i_1 \ldots i_p})$, where $t_{j_1 \ldots j_q}^{i_1 \ldots i_p}$ are the components of t with respect to the natural basis in $T_q^p(P)$. We may consider

$$(x^j, t_{j_1 \ldots j_q}^{i_1 \ldots i_p}) = (x^j, x^{\bar{j}}) = (x^J),\ j = 1, \ldots, n,\ \bar{j} = n+1, \ldots, n + n^{p+q},$$
$$J = 1, \ldots, n + n^{p+q}.$$

as local coordinates in a neighborhood $\pi^{-1}(U) \subset T_q^p(M_n)$.

It is straightforward to see that $T_q^p(M_n)$ becomes an $(n + n^{p+q})$-manifold; indeed if $x^{j'}$, $j' = 1, \ldots, n$ are local coordinates in another neighborhood V of $P \in M_n$, with $U \cap V \neq \emptyset$, then the change of coordinates is given by

$$\begin{cases} x^{j'} = x^{j'}(x^j), \\ x^{\bar{j}'} = t_{j'_1 \ldots j'_q}^{i'_1 \ldots i'_p} = A_{i_1}^{i'_1} \ldots A_{i_p}^{i'_p} A_{j'_1}^{j_1} \ldots A_{j'_q}^{j_q} t_{j_1 \ldots j_q}^{i_1 \ldots i_p} = A_{(i)}^{(i')} A_{(j')}^{(j)} x^{\bar{j}}, \end{cases} \tag{3.1}$$

where

$$A_{(i)}^{(i')} A_{(j')}^{(j)} = A_{i_1}^{i'_1} \ldots A_{i_p}^{i'_p} A_{j'_1}^{j_1} \ldots A_{j'_q}^{j_q},\ A_i^{i'} = \frac{\partial x^{i'}}{\partial x^i},\ A_{j'}^{j} = \frac{\partial x^j}{\partial x^{j'}}.$$

The Jacobian of (3.1) is:

$$\left(\frac{\partial x^{J'}}{\partial x^J} \right) = \begin{pmatrix} \frac{\partial x^{j'}}{\partial x^j} & \frac{\partial x^{j'}}{\partial x^{\bar{j}}} \\ \frac{\partial x^{\bar{j}'}}{\partial x^j} & \frac{\partial x^{\bar{j}'}}{\partial x^{\bar{j}}} \end{pmatrix} = \begin{pmatrix} A_j^{j'} & 0 \\ t_{(k)}^{(i)} \partial_j (A_{(i)}^{(i')} A_{(j')}^{(k)}) & A_{(i)}^{(i')} A_{(j')}^{(j)} \end{pmatrix}, \tag{3.2}$$

where $J = (j, \bar{j})$, $J = 1, \ldots, n + n^{p+q}$, $t_{(k)}^{(i)} = t_{k_1 \ldots k_q}^{i_1 \ldots i_p}$.

We denote by $\Im_s^r(M_n)$ the module over $F(M_n)$ ($F(M_n)$ is the ring of C^∞-functions on M_n) all tensor fields of class C^∞ and of type (r, s) on M_n. If $\alpha \in \Im_p^q(M_n)$, it is regarded, in a natural way (by contraction), as a function in $T_q^p(M_n)$, which we denote by $\alpha(t) = \iota\alpha$. If α has the local expression

$$\alpha = \alpha_{i_1 \ldots i_p}^{j_1 \ldots j_q} \partial_{j_1} \otimes \cdots \otimes \partial_{j_q} \otimes dx^{i_1} \otimes \cdots \otimes dx^{i_p}$$

in a coordinate neighborhood $U(x^j) \subset M_n$, then $\alpha(t) = \iota\alpha$ has the local expression

$$\iota\alpha = \alpha^{j_1 \dots j_q}_{i_1 \dots i_p} t^{i_1 \dots i_p}_{j_1 \dots j_q}$$

with respect to the coordinates $(x^j, x^{\overline{j}})$ in $\pi^{-1}(U)$.

For later use, we first shall state following theorem [9]:

Theorem 3.1 *Let $\tilde{X}$ and $\tilde{Y}$ be vector fields on $T^p_q(M_n)$ such that $\tilde{X}(\iota\alpha) = \tilde{Y}(\iota\alpha)$, for any $\alpha \in \Im^q_p(M)$. Then $\tilde{X} = \tilde{Y}$, i.e. $\tilde{X} \in \Im^1_0(T^p_q(M))$ is completely determined by its action on functions of type $\iota\alpha$.*

Suppose that $A \in \Im^p_q(M_n)$. Then there is a unique vector field ${}^V A \in \Im^1_0(T^p_q(M_n))$ such that for $\alpha \in \Im^q_p(M_n)$ [30]

$$ {}^V A(\iota\alpha) = \alpha(A) \circ \pi = {}^V(\alpha(A)), \tag{3.3}$$

where ${}^V(\alpha(A))$ is the vertical lift of the function $\alpha(A) \in F(M_n)$. We note that the vertical lift ${}^V f = f \circ \pi$ of an arbitrary function $f \in F(M_n)$ is constant along each fibre $\pi^{-1}(P)$. The vector field ${}^V A$ is called the *vertical lift of* A to $T^p_q(M_n)$. If ${}^V A = {}^V A^k \partial_k + {}^V A^{\overline{k}} \partial_{\overline{k}}$, then from (3.3) we see that the vertical lift ${}^V A$ of A to $T^p_q(M_n)$ has the components

$$ {}^V A = \begin{pmatrix} {}^V A^j \\ {}^V A^{\overline{j}} \end{pmatrix} = \begin{pmatrix} 0 \\ A^{i_1 \dots i_p}_{j_1 \dots j_q} \end{pmatrix} \tag{3.4}$$

with respect to the coordinates $(x^j, x^{\overline{j}})$ in $T^p_q(M_n)$.

Let now $\varphi \in \Im^1_1(M_n)$, i.e.

$$\varphi = \varphi^i_j \frac{\partial}{\partial x^i} \otimes dx^j.$$

Two new vector fields $\gamma\varphi \in \Im^1_0(T^p_q(M_n))$, $\tilde{\gamma}\varphi \in \Im^1_0(T^p_q(M_n))$ are defined by

$$\begin{cases} \gamma\varphi = (\sum\limits_{\lambda=1}^{p} t^{i_1 \dots m \dots i_p}_{j_1 \dots j_q} \varphi^{i_\lambda}_m) \dfrac{\partial}{\partial x^{\overline{j}}}, & (p \geq 1, q \geq 0), \\ \tilde{\gamma}\varphi = (\sum\limits_{\mu=1}^{q} t^{i_1 \dots i_p}_{j_1 \dots m \dots j_q} \varphi^m_{j_\mu}) \dfrac{\partial}{\partial x^{\overline{j}}}, & (p \geq 0, q \geq 1), \end{cases}$$

with respect to the coordinates $(x^j, x^{\overline{j}})$ in $T^p_q(M_n)$. From (3.2) we easily see that the vector fields $\gamma\varphi$ and $\tilde{\gamma}\varphi$ determine respectively global vector fields on $T^p_q(M_n)$.

Let L_V be the Lie derivative with respect to a vector field $V \in \Im^1_0(M_n)$. The *complete lift* ${}^c V$ of V to $T^p_q(M_n)$ is defined by [30]

$$^{c}V(\iota\alpha) = \iota(L_V\alpha), \tag{3.5}$$

for any $\alpha \in \Im_p^q(M)$. If $^{c}V = {}^{c}V^k\partial_k + {}^{c}V^{\overline{k}}\partial_{\overline{k}}$, then from (3.5) we see that the complete lift ^{c}V of V has components

$$^{c}V = \begin{pmatrix} ^{c}V^j \\ ^{c}V^{\overline{j}} \end{pmatrix} = \begin{pmatrix} V^j \\ \sum\limits_{\lambda=1}^{p} t_{j_1 \ldots j_q}^{i_1 \ldots m \ldots i_p} \partial_m V^{i_\lambda} - \sum\limits_{\mu=1}^{q} t_{j_1 \ldots m \ldots j_q}^{i_1 \ldots i_p} \partial_{j_\mu} V^m \end{pmatrix} \tag{3.6}$$

with respect to the coordinates $(x^j, x^{\overline{j}})$ in $T_q^p(M)$ (see [31, 48, 49, 61, 62, 68]).

Let now ∇_V be a covariant derivative with respect to the vector field $V \in \Im_1^0(M_n)$, where ∇ is the symmetric affine connection on M_n. We define the *horizontal lift* $^{H}V \in \Im_0^1(T_q^p(M_n))$ of $V \in \Im_1^0(M_n)$ to $T_q^p(M_n)$ by [30]

$$^{H}V(\iota\alpha) = \iota(\nabla_V\alpha), \quad \forall\alpha \in \Im_p^q(M_n).$$

From here we see that the horizontal lift ^{H}V of $V \in \Im_1^0(M_n)$ to $T_q^p(M_n)$ has components

$$^{H}V = \begin{pmatrix} V^j \\ V^s\Big(\sum\limits_{\mu=1}^{q} \Gamma_{sj_\mu}^{m} t_{j_1 \ldots m \ldots j_q}^{i_1 \ldots i_p} - \sum\limits_{\lambda=1}^{p} \Gamma_{sm}^{i_\lambda} t_{j_1 \ldots j_q}^{i_1 \ldots m \ldots i_p}\Big) \end{pmatrix}$$

with respect to the coordinates $(x^j, x^{\overline{j}})$ in $T_q^p(M_n)$, where Γ_{ij}^k are local components of ∇ (see [49]).

Suppose that there is given a tensor field $\xi \in \Im_q^p(M_n)$. If the correspondence $x \to \xi_x$, where ξ_x is the value of ξ at $x \in M_n$, determines a mapping $\sigma_\xi : M_n \to T_q^p(M_n)$ such that $\pi \circ \sigma_\xi = id_{M_n}$, then the n dimensional submanifold $\sigma_\xi(M_n)$ of $T_q^p(M_n)$ is called the *cross-section* determined by ξ. If the tensor field ξ has the local components $\xi_{k_1 \ldots k_q}^{h_1 \ldots h_p}(x^k)$, the cross-section $\sigma_\xi(M_n)$ is locally expressed by

$$\begin{cases} x^k = x^k, \\ x^{\overline{k}} = \xi_{k_1 \ldots k_q}^{h_1 \ldots h_p}(x^k). \end{cases} \tag{3.7}$$

Differentiating (3.7) by x^j, we see that n tangent vector fields B_j to $\sigma_\xi(M_n)$ have the components

$$(B_j^K) = \left(\frac{\partial x^K}{\partial x^j}\right) = \begin{pmatrix} \delta_j^k \\ \partial_j \xi_{k_1 \ldots k_q}^{h_1 \ldots h_p} \end{pmatrix} \tag{3.8}$$

with respect to the natural frame $\{\partial_k, \partial_{\overline{k}}\}$ in $T_q^p(M_n)$.

On the other hand, the fibre is locally expressed by

$$\begin{cases} x^k = \text{const}, \\ x^{\overline{k}} = t_{k_1 \dots k_q}^{h_1 \dots h_p}, \end{cases}$$

where $t_{k_1 \dots k_q}^{h_1 \dots h_p}$ are considered as parameters. Thus, by differentiating with respect to $x^{\overline{j}} = t_{j_1 \dots j_q}^{i_1 \dots i_p}$, we see that n^{p+q} tangent vector fields $C_{\overline{j}}$ to the fibre have components

$$(C_{\overline{j}}^K) = \left(\frac{\partial x^K}{\partial x^{\overline{j}}}\right) = \begin{pmatrix} 0 \\ \delta_{k_1}^{j_1} \dots \delta_{k_q}^{j_q} \delta_{i_1}^{h_1} \dots \delta_{i_p}^{h_p} \end{pmatrix} \tag{3.9}$$

with respect to the natural frame $\{\partial_k, \partial_{\overline{k}}\}$ on $T_q^p(M_n)$, where δ is the Kronecker symbol.

We consider in $\pi^{-1}(U) \subset T_q^p(M_n)$, $n + n^{p+q}$ local vector fields B_j and $C_{\overline{j}}$ along $\sigma_\xi(M_n)$. They form a family of local frames $\{B_j, C_{\overline{j}}\}$ along $\sigma_\xi(M_n)$, which is called the adapted (B, C)-frame of $\sigma_\xi(M_n)$ in $\pi^{-1}(U)$. From ${}^cV = {}^cV^h\partial_h + {}^cV^{\overline{h}}\partial_{\overline{h}}$ and ${}^cV = {}^c\overline{V}^j B_j + {}^c\overline{V}^{\overline{j}} C_{\overline{j}}$ we easily obtain ${}^cV^k = {}^c\overline{V}^j B_j^k + {}^c\overline{V}^{\overline{j}} C_{\overline{j}}^k$, ${}^cV^{\overline{k}} = {}^c\overline{V}^j B_j^{\overline{k}} + {}^c\overline{V}^{\overline{j}} C_{\overline{j}}^{\overline{k}}$. Now, taking account of (3.6) on the cross-section $\sigma_\xi(M_n)$, and of (3.8) and (3.9) also, we firstly find ${}^c\overline{V}^k = V^k$ and therefore ${}^c\overline{V}^{\overline{k}} = -L_V \xi_{k_1 \dots k_q}^{h_1 \dots h_p}$. Thus, the complete lift ${}^cV$ has along $\sigma_\xi(M_n)$ components of the form

$${}^cV = \begin{pmatrix} {}^c\overline{V}^k \\ {}^c\overline{V}^{\overline{k}} \end{pmatrix} = \begin{pmatrix} V^k \\ -L_V \xi_{k_1 \dots k_q}^{h_1 \dots h_p} \end{pmatrix} \tag{3.10}$$

with respect to the adapted (B, C)-frame.

From (3.4), (3.8) and (3.9) we see that the vertical lift ${}^VA$ also has components

$${}^VA = \begin{pmatrix} {}^V\bar{A}^k \\ {}^V\bar{A}^{\overline{k}} \end{pmatrix} = \begin{pmatrix} 0 \\ A_{k_1 \dots k_q}^{h_1 \dots h_p} \end{pmatrix} \tag{3.11}$$

with respect to the adapted (B, C)-frame.

In a similar way we see that the horizontal lift ${}^HV$ has along $\sigma_\xi(M_n)$ components of the form

$${}^HV = \begin{pmatrix} {}^H\bar{V}^j \\ {}^H\bar{V}^{\overline{j}} \end{pmatrix} = \begin{pmatrix} V^j \\ -(\nabla_V \xi)_{j_1 \dots j_q}^{i_1 \dots i_p} \end{pmatrix}$$

with respect to the adapted (B, C)-frame, where $(\nabla_V \xi)_{j_1 \dots j_q}^{i_1 \dots i_p}$ are local components of $\nabla_V \xi$.

Let now $\varphi \in \mathfrak{I}_1^1(M)$. We can easily verify that $\gamma\varphi$ and $\tilde{\gamma}\varphi$ have along $\sigma_\xi(M)$ components

$$\gamma\varphi = ((\overline{\gamma\varphi})^J) = \begin{pmatrix} 0 \\ \sum_{\lambda=1}^{p} \xi_{j_1 \dots j_q}^{i_1 \dots m \dots i_p} \varphi_m^{i_\lambda} \end{pmatrix}, \quad \tilde{\gamma}\varphi = ((\overline{\tilde{\gamma}\varphi})^J) = \begin{pmatrix} 0 \\ \sum_{\mu=1}^{q} \xi_{j_1 \dots m \dots j_q}^{i_1 \dots i_p} \varphi_{j_\mu}^m \end{pmatrix} \tag{3.12}$$

with respect to the adapted (B, C)-frame.

3.2 Complete Lifts of (1,1)-Tensor Fields

Let $\varphi \in \Im_1^1(M_n)$. The cross-section $\sigma_\xi : M_n \to T_q^p(M_n)$ defined by the tensor field $\xi \in \Im_q^p(M_n)$ is called a *pure cross-section* if ξ is a pure tensor field with respect to φ (See Chap. 1) and denoted by $\sigma_\xi^\varphi(M_n)$. The complete lift ${}^c\varphi$ of φ along the pure cross-section of $T_q^p(M_n)$ is defined as

Definition 3.1 [31, 50, 68] We define a tensor field ${}^c\varphi \in \Im_1^1(T_q^p(M_n))$ along the pure cross-section $\sigma_\xi^\varphi(M_n)$ by

$$\begin{cases} {}^c\varphi({}^cV) = {}^c(\varphi(V)) - \gamma(L_V\varphi) + {}^V((L_V\varphi)\circ\xi),\ \forall V \in \Im_0^1(M_n), & (i) \\ {}^c\varphi({}^VA) = {}^V(\varphi(A)),\ \forall A \in \Im_q^p(M_n), & (ii) \end{cases} \tag{3.13}$$

and call ${}^c\varphi$ the complete lift of $\varphi \in \Im_1^1(M_n)$ to $T_q^p(M_n)$, $p \geq 1, q \geq 0$ along $\sigma_\xi^\varphi(M_n)$, where $\varphi(A) \in \Im_q^p(M_n)$, $((L_V\varphi)\circ\xi)(x_1, \dots, x_q, \alpha_1, \dots, \alpha_p) \in \Im_q^p(M_n)$.

In particular, if we assume that $p = 1$, $q \geq 0$, then from (3.11) and (3.12) we get

$$\gamma(L_V\varphi) = {}^V((L_V\varphi)\circ\xi).$$

Substituting this into (3.13), we find

$${}^c\varphi({}^cV) = {}^c(\varphi(V)),\ {}^c\varphi({}^VA) = {}^V(\varphi(A))\ .$$

We observe that the local vector fields

$${}^cX_{(j)} = {}^c\left(\frac{\partial}{\partial x^j}\right) = {}^c\left(\delta_j^h \frac{\partial}{\partial x^h}\right) = \begin{pmatrix} \delta_j^h \\ 0 \end{pmatrix}$$

and

$$\begin{aligned} {}^VX^{(\bar{j})} &= {}^V(\partial_{j_1} \otimes \dots \otimes \partial_{j_p} dx^{i_1} \otimes \dots \otimes dx^{i_q}) \\ &= {}^V(\delta_{h_1}^{i_1} \dots \delta_{h_q}^{i_q} \delta_{j_1}^{k_1} \dots \delta_{j_p}^{k_p} \partial_{k_1} \otimes \dots \otimes \partial_{k_p} \otimes dx^{h_1} \otimes \dots \otimes dx^{h_q}) \end{aligned}$$

$$= \begin{pmatrix} 0 \\ \delta^{i_1}_{h_1} \ldots \delta^{i_q}_{h_q} \delta^{k_1}_{j_1} \ldots \delta^{k_p}_{j_p} \end{pmatrix},$$

$j = 1, \ldots, n$, $\overline{j} = n+1, \ldots, n+n^{p+q}$ span the module of all vector fields in $\pi^{-1}(U)$. Hence any tensor field is determined in $\pi^{-1}(U)$ by its action on ${}^cX_{(j)}$ and ${}^VX^{(\overline{j})}$.

Now, let us calculate the components of the complete lift of tensor fields of type (1,1) by using the Φ_φ-operator.

Theorem 3.1 *Let $\varphi \in \Im^1_1(M_n)$ and $\sigma^\varphi_\xi(M_n)$ be a pure cross-section of $T^p_q(M_n)$ with respect to φ. Then the complete lift ${}^C\varphi \in \Im^1_1(T^p_q(M_n))$ of φ has along the pure cross-section $\sigma^\varphi_\xi(M_n)$ the components*

$$\begin{cases} {}^C\overline{\varphi}^k_l = \varphi^k_l, \ {}^C\overline{\varphi}^k_{\overline{l}} = 0, \ {}^C\overline{\varphi}^{\overline{k}}_l = -(\Phi_\varphi \xi)^{h_1 \ldots h_p}_{l k_1 \ldots k_q}, \\ {}^C\overline{\varphi}^{\overline{k}}_{\overline{l}} = \varphi^{h_1}_{s_1} \delta^{h_2}_{s_2} \ldots \delta^{h_p}_{s_p} \delta^{r_1}_{k_1} \ldots \delta^{r_q}_{k_q} \end{cases} \tag{3.14}$$

with respect to the adapted (B, C)-frame of $\sigma^\varphi_\xi(M_n)$, where Φ_φ is the Tachibana operator and $x^{\overline{k}} = t^{h_1 \ldots h_p}_{k_1 \ldots k_q}$, $x^{\overline{l}} = t^{s_1 \ldots s_p}_{r_1 \ldots r_q}$.

Remark 3.1 The formula (3.14) is a useful expansion of (3.10) to (1,1)-tensor fields along the pure cross-section by using the Tachibana operator instead of the Lie derivative.

Proof Let ${}^c\overline{\varphi}^K_L$ be components of ${}^c\varphi$ with respect to the adapted (B, C)-frame of the pure cross-section $\sigma^\varphi_\xi(M_n)$. Then from (3.13) we have

$$\begin{cases} {}^c\bar{\varphi}^K_L {}^c\bar{V}^L = {}^c(\overline{\varphi(V)})^K - (\overline{\gamma(L_V\varphi)})^K + {}^V(\overline{(L_V\varphi) \circ \xi})^K, & (i) \\ {}^c\bar{\varphi}^K_L {}^V\bar{A}^L = {}^V(\overline{\varphi(A)})^K, & (ii) \end{cases} \tag{3.15}$$

where

$$({}^V(\overline{\varphi(A)})^K) = \begin{pmatrix} 0 \\ \varphi^{h_1}_m A^{m h_2 \ldots h_p}_{k_1 \ldots k_q} \end{pmatrix}, \ {}^V(\overline{(L_V\varphi) \circ \xi})^K = \begin{pmatrix} 0 \\ (L_V \varphi^{h_\lambda}_m) \xi^{h_1 \ldots m \ldots h_p}_{k_1 \ldots k_q} \end{pmatrix},$$

$$(\overline{\gamma(L_V\varphi)})^K = \begin{pmatrix} 0 \\ ((L_V\varphi)^{h_1}_m) \xi^{m h_2 \ldots h_p}_{k_1 \ldots k_q} \end{pmatrix}.$$

First, consider the case where $K = k$. In this case, (i) of (3.15) reduces to

$${}^c\overline{\varphi}^k_l {}^c\overline{V}^l + {}^c\overline{\varphi}^k_{\overline{l}} {}^c\overline{V}^{\overline{l}} = {}^c(\overline{\varphi(V)})^k = (\varphi(V))^k = \varphi^k_l V^l. \tag{3.16}$$

Since the right-hand side of (3.16) are functions depending only on the base coordinates x^i, the left-hand sides of (3.16) are too. Then, since ${}^cV^{\overline{l}} \neq 0$ depend on fibre coordinates, from (3.16) we obtain

$$^{c}\overline{\varphi}_{\overline{l}}^{k} = 0. \tag{3.17}$$

From (3.16) and (3.17) we have $^{c}\overline{\varphi}_{l}^{k}{}^{c}\overline{V}^{l} = {}^{c}\overline{\varphi}_{l}^{k} V^{l} = \varphi_{l}^{k} V^{l}$, V^{i} being arbitrary, which implies

$$^{c}\overline{\varphi}_{l}^{k} = \varphi_{l}^{k}.$$

When $K = \overline{k}$, (ii) of (3.15) reduces to

$$^{c}\overline{\varphi}_{l}^{\overline{k}}\, {}^{V}\overline{A}^{l} + {}^{c}\overline{\varphi}_{\overline{l}}^{\overline{k}}\, {}^{V}\overline{A}^{\overline{l}} = {}^{V}\overline{(\varphi(A))}^{\overline{k}}$$

or

$$^{c}\overline{\varphi}_{\overline{l}}^{\overline{k}} A_{r_1 \ldots r_q}^{s_1 \ldots s_p} = \varphi_{m}^{h_1} A_{k_1 \ldots k_q}^{m h_2 \ldots h_p} = \delta_{k_1}^{r_1} \ldots \delta_{k_q}^{r_q} \varphi_{s_1}^{h_1} \delta_{s_2}^{h_2} \ldots \delta_{s_p}^{h_p} A_{r_1 r_2 \ldots r_q}^{s_1 s_2 \ldots s_p}$$

for all $A \in \Im_q^p(M_n)$, which implies

$$^{c}\overline{\varphi}_{\overline{l}}^{\overline{k}} = \delta_{k_1}^{r_1} \ldots \delta_{k_q}^{r_q} \varphi_{s_1}^{h_1} \delta_{s_2}^{h_2} \ldots \delta_{s_p}^{h_p},$$

where $x^{\overline{k}} = t_{k_1 \ldots k_q}^{h_1 \ldots h_p}$, $x^{\overline{l}} = t_{r_1 \ldots r_q}^{s_1 \ldots s_p}$.

When $K = \overline{k}$, (i) of (3.15) reduces to

$$^{c}\overline{\varphi}_{l}^{\overline{k}}{}^{c}\overline{V}^{l} + {}^{c}\overline{\varphi}_{\overline{l}}^{\overline{k}}{}^{c}\overline{V}^{\overline{l}} = {}^{c}\overline{(\varphi(V))}^{\overline{k}} - \sum_{\lambda=2}^{p} (L_V \varphi_{l}^{h_\lambda}) \xi_{k_1 \ldots k_q}^{h_1 \ldots l \ldots h_p}$$

or

$$^{c}\overline{\varphi}_{l}^{\overline{k}}{}^{c}\overline{V}^{l} + \varphi_{s_1}^{h_1} \delta_{s_2}^{h_2} \ldots \delta_{s_p}^{h_p} \delta_{k_1}^{r_1} \ldots \delta_{k_q}^{r_q}{}^{c}\overline{V}^{\overline{l}} + \sum_{\lambda=2}^{p} (L_V \varphi_{l}^{h_\lambda}) \xi_{k_1 \ldots k_q}^{h_1 h_2 \ldots l \ldots h_p} = {}^{c}\overline{(\varphi(V))}^{\overline{k}}. \tag{3.18}$$

Now, using the Tachibana operator we will investigate the components $^{c}\overline{\varphi}_{l}^{\overline{k}}$. Let $\overset{*}{\Im}{}_q^p(M_n)$ denote module of all the tensor fields $\xi \in \Im_q^p(M_n)$ which are pure with respect to φ. The Tachibana operator on the pure module $\overset{*}{\Im}{}_q^p(M_n)$ is given by (see Chap. 1)

$$(\Phi_\varphi \xi)_{l k_1 \ldots k_q}^{h_1 \ldots h_p} = \varphi_{l}^{m} \partial_m \xi_{k_1 \ldots k_q}^{h_1 \ldots h_p} - \partial_l \overset{*}{\xi}{}_{k_1 \ldots k_q}^{h_1 \ldots h_p} + \sum_{a=1}^{q} (\partial_{k_a} \varphi_{l}^{m}) \xi_{k_1 \ldots m \ldots k_q}^{h_1 \ldots h_p} + 2 \sum_{\lambda=1}^{p} \partial_{[l} \varphi_{m]}^{h_\lambda} \xi_{k_1 \ldots k_q}^{h_1 \ldots m \ldots h_p},$$

where

$$\varphi_{k_1}^{m} \xi_{m k_2 \ldots k_q}^{h_1 \ldots h_p} = \varphi_{k_2}^{m} \xi_{k_1 m \ldots k_q}^{h_1 \ldots h_p} = \cdots = \varphi_{k_q}^{m} \xi_{k_1 k_2 \ldots m}^{h_1 \ldots h_p} = \varphi_{m}^{h_1} \xi_{k_1 \ldots k_q}^{m h_2 \ldots h_p} = \varphi_{m}^{h_2} \xi_{k_1 \ldots k_q}^{h_1 m \ldots h_p}$$

$$= \cdots = \varphi_m^{h_p} \xi_{k_1 \ldots k_q}^{h_1 h_2 \ldots m} = \overset{*}{\xi}\,{}_{k_1 k_2 \ldots k_q}^{h_1 h_2 \ldots h_p}.$$

After some calculations we have

$$V^l (\Phi_\varphi \xi)_{l k_1 \ldots k_q}^{h_1 \ldots h_p} = (L_{\varphi V} \xi)_{k_1 \ldots k_q}^{h_1 \ldots h_p} - (L_V \overset{*}{\xi})_{k_1 \ldots k_q}^{h_1 \ldots h_p} + \sum_{\lambda=1}^{p} (L_V \varphi_m^{h_\lambda}) \xi_{k_1 \ldots k_q}^{h_1 \ldots m \ldots h_p}$$

or

$$V^l (\Phi_\varphi \xi)_{l k_1 \ldots k_q}^{h_1 \ldots h_p} + \varphi_m^{h_1} (L_V \xi)_{k_1 \ldots k_q}^{m h_2 \ldots h_p} + \xi_{k_1 \ldots k_q}^{m h_2 \ldots h_p} (L_V \varphi)_m^{h_1} - \sum_{\lambda=1}^{p} (L_V \varphi_m^{h_\lambda}) \xi_{k_1 \ldots k_q}^{h_1 \ldots m \ldots h_p} = (L_{\varphi V} \xi)_{k_1 \ldots k_q}^{h_1 \ldots h_p} \tag{3.19}$$

for any $V \in \Im_0^1(M_n)$. Using (3.10), from (3.19) we have

$$\begin{aligned}
& V^l (\Phi_\varphi \xi)_{l k_1 \ldots k_q}^{h_1 \ldots h_p} + \varphi_m^{h_1} (L_V \xi)_{k_1 \ldots k_q}^{m h_2 \ldots h_p} + \xi_{k_1 \ldots k_q}^{m h_2 \ldots h_p} (L_V \varphi_m^{h_1}) - \sum_{\lambda=1}^{p} (L_V \varphi_m^{h_\lambda}) \xi_{k_1 \ldots k_q}^{h_1 \ldots m \ldots h_p} \\
&= V^l (\Phi_\varphi \xi)_{l k_1 \ldots k_q}^{h_1 \ldots h_p} + \varphi_{s_1}^{h_1} \delta_{s_2}^{h_2} \ldots \delta_{s_p}^{h_p} \delta_{k_1}^{r_1} \ldots \delta_{k_q}^{r_q} (L_V \xi)_{r_1 \ldots r_q}^{s_1 \ldots s_p} - \sum_{\lambda=2}^{p} (L_V \varphi_m^{h_\lambda})_{k_1 \ldots k_q}^{h_1 h_2 \ldots m \ldots h_p} \\
&= {}^c V^l (\Phi_\varphi \xi)_{l k_1 \ldots k_q}^{h_1 \ldots h_p} - \varphi_{s_1}^{h_1} \delta_{s_2}^{h_2} \ldots \delta_{s_p}^{h_p} \delta_{k_1}^{r_1} \ldots \delta_{k_q}^{r_q}\, {}^c V^{\bar{l}} - \sum_{\lambda=2}^{p} (L_V \varphi_m^{h_\lambda}) \xi_{k_1 \ldots k_q}^{h_1 h_2 \ldots m \ldots h_p} \\
&= -{}^c(\varphi(V))^{\bar{k}}
\end{aligned}$$

or

$$-(\Phi_\varphi \xi)_{l k_1 \ldots k_q}^{h_1 \ldots h_p}\, {}^c V^l + \varphi_{s_1}^{h_1} \delta_{s_2}^{h_2} \ldots \delta_{s_p}^{h_p} \delta_{k_1}^{r_1} \ldots \delta_{k_q}^{r_q}\, {}^c V^{\bar{l}} + \sum_{\lambda=2}^{p} (L_V \varphi_m^{h_\lambda}) \xi_{k_1 \ldots k_q}^{h_1 h_2 \ldots m \ldots h_p} = {}^c(\varphi(V))^{\bar{k}}. \tag{3.20}$$

Comparing (3.18) and (3.20), we get

$${}^c \overline{\varphi}_l^{\bar{k}} = -(\Phi_\varphi \xi)_{l k_1 \ldots k_q}^{h_1 \ldots h_p}.$$

This completes the proof.

Remark 3.2 We note that the lift ${}^c\varphi$ in the form (3.14) is the unique solution of (3.13). Therefore, if $\overset{*}{\varphi}$ is another element of $\Im_1^1(T_q^p(M_n))$ such that

$$\overset{*}{\varphi}({}^c V) = {}^c\varphi({}^c V) = {}^c(\varphi(V)) - \gamma(L_V \varphi) + {}^V((L_V \varphi) \circ \xi), \; \overset{*}{\varphi}({}^V A) = {}^c\varphi({}^V A) = {}^V(\varphi(A)),$$

then $\overset{*}{\varphi} = {}^c\varphi$.

Remark 3.3 If $p = 1, q = 0$, then (3.14) is the formula of the complete lift of affinor fields to the tangent bundle along the cross-section $\sigma_\xi^\varphi(M_n)$ (for details, see [87, p. 126]).

Now, setting $B_{\overline{j}} = C_{\overline{j}}$ we write the adapted (B, C)-frame of $\sigma_\xi^\varphi(M_n)$ as $B_J = \left\{B_j, B_{\overline{j}}\right\}$. We define a coframe $\tilde{B}^J$ of $\sigma_\xi^\varphi(M_n)$ by $\tilde{B}^I(B_J) = \delta_J^I$. From (3.8), (3.9) and $B_J^K \tilde{B}_K^I = \delta_J^I$ we see that covector fields $\tilde{B}^I$ have components

$$\begin{aligned} \tilde{B}^i &= (\tilde{B}_K^i) = (\delta_k^i, 0), \\ \tilde{B}^{\overline{i}} &= (\tilde{B}_K^{\overline{i}}) = (-\partial_k \xi_{i_1 \ldots i_q}^{j_1 \ldots j_p}, \delta_{i_1}^{k_1} \ldots \delta_{i_q}^{k_q} \delta_{h_1}^{j_1} \ldots \delta_{h_p}^{j_p}) \end{aligned} \tag{3.21}$$

with respect to the natural coframe $(\mathrm{d}x^k, \mathrm{d}x^{\overline{k}})$. Taking account of

$$\begin{aligned} {}^c\varphi_L^K &= {}^c\varphi(\mathrm{d}x^K, \partial_L) = {}^c\overline{\varphi}_I^J B_J \otimes \tilde{B}^I(\mathrm{d}x^K, \partial_L) = {}^c\overline{\varphi}_I^J \mathrm{d}x^K(B_J)\tilde{B}^I(\partial_L) \\ &= {}^c\overline{\varphi}_I^J \mathrm{d}x^K(B_J^H \partial_H)\tilde{B}_L^I = {}^c\overline{\varphi}_I^J B_J^H \delta_H^K \tilde{B}_L^I = {}^c\overline{\varphi}_I^J B_J^K \tilde{B}_L^I, \end{aligned}$$

and also (3.8), (3.9), (3.14) and (3.21), we see that the complete lift ${}^c\varphi$ has along the pure cross-section $\sigma_\xi^\varphi(M_n)$ components of the form

$${}^c\varphi_l^k = \varphi_l^k, \ {}^c\varphi_{\overline{l}}^k = 0, \ {}^c\varphi_{\overline{l}}^{\overline{k}} = \varphi_{s_1}^{h_1} \delta_{s_2}^{h_2} \ldots \delta_{s_p}^{h_p} \delta_{k_1}^{r_1} \ldots \delta_{k_q}^{r_q},$$

$${}^c\varphi_l^{\overline{k}} = (\partial_l \varphi_m^{h_1}) \xi_{k_1 \ldots k_q}^{m h_2 \ldots h_p} - \sum_{\mu=1}^{q} (\partial_{k_\mu} \varphi_l^m) \xi_{k_1 \ldots m \ldots k_q}^{h_1 \ldots h_p} - \sum_{\lambda=1}^{p} (\partial_l \varphi_m^{h_\lambda} - \partial_m \varphi_l^{h_\lambda}) \xi_{k_1 \ldots k_q}^{h_1 \ldots m \ldots h_p}$$

with respect to the natural frame $\{\partial_k, \partial_{\overline{k}}\}$ of $\sigma_\xi^\varphi(M_n)$ in $\pi^{-1}(U)$ [70].

3.3 Holomorphic Cross-Sections

Theorem 3.2 *If φ is an integrable almost complex structure on M_n, then the complete lift ${}^C\varphi$ of φ to $T_q^p(M_n)$, $p \geq 1$, $q \geq 0$ along the pure cross-section $\sigma_\xi^\varphi(M_n)$ is an almost complex structure.*

Proof Let $\varphi \in \Im_1^1(M_n)$ and $S \in \Im_2^1(M_n)$. Using (3.10), (3.12) and (3.14) we have

$$\gamma(\varphi \pm \psi) = \gamma\varphi \pm \gamma\psi, \ {}^C\varphi(\gamma\psi) = \gamma(\varphi \circ \psi) = \gamma(\psi \circ \varphi), (\gamma S)^C V = \gamma S_V, \tag{3.22}$$

where S_V is the tensor field of type (1.1) on M_n defined by $S_V(W) = S(V, W)$ for any $W \in \Im_0^1(M_n)$ and γS is the affinor field along $\sigma_\xi^\varphi(M_n)$ with the components

$$\gamma S = \begin{pmatrix} 0 & 0 \\ \sum\limits_{\lambda=1}^{p} S_{jm}^{j_\lambda} \xi_{i_1 \ldots i_q}^{j_1 \ldots m \ldots j_p} & 0 \end{pmatrix} \tag{3.23}$$

with respect to the adapted (B, C)-frame, $S_{jm}^{j_1}$ are the local components of S. It is clear that ${}^V(S \circ \xi)$ is also affinor field along $\sigma_\xi(M_n)$ with the components

$${}^V(S \circ \xi) = \begin{pmatrix} 0 & 0 \\ S_{jm}^{j_1} \xi_{i_1 \ldots i_q}^{m j_2 \ldots j_p} & 0 \end{pmatrix}$$

with respect to the adapted (B, C)-frame.

If $V \in \Im_0^1(M_n)$, then from (3.13) and (3.22) we have

$$\begin{aligned}
({}^C\varphi)^2({}^CV) &= ({}^C\varphi \circ {}^C\varphi){}^CV = {}^C\varphi({}^C\varphi({}^CV)) = {}^C\varphi({}^C(\varphi(V))) - \gamma(L_V\varphi) + {}^V((L_V\varphi)\circ\xi) \\
&= {}^C\varphi({}^C(\varphi(V))) - {}^C\varphi(\gamma(L_V\varphi)) + {}^C\varphi({}^V((L_V\varphi)\circ\xi)) \\
&= {}^C(\varphi(\varphi(V))) - \gamma(L_{\varphi(V)}\varphi) - {}^C\varphi(\gamma(L_V\varphi)) + {}^C\varphi({}^V((L_V\varphi)\circ\xi)) + {}^V((L_{\varphi(V)}\varphi)\circ\xi) \\
&= {}^C((\varphi\circ\varphi)(V)) - \gamma(L_{\varphi(V)}\varphi) - \gamma((L_V\varphi)\circ\varphi) + {}^V(\varphi((L_V\varphi)\circ\xi) + (L_{\varphi(V)}\varphi)\circ\xi) \\
&= {}^C((\varphi\circ\varphi)(V)) - \gamma(L_{\varphi(V)}\varphi + (L_V\varphi)\circ\varphi) + {}^V((L_{\varphi(V)}\varphi) + (L_V\varphi)\circ\xi) \\
&\quad + {}^V((L_V(\varphi\circ\varphi))\circ\xi) = {}^C(\varphi\circ\varphi)({}^CV) + \gamma(L_V(\varphi\circ\varphi)) - \gamma(L_{\varphi(V)}\varphi + (L_V\varphi)\circ\varphi) \\
&= {}^C(\varphi\circ\varphi)({}^CV) - \gamma(L_{\varphi(V)}\varphi - \varphi\circ(L_V\varphi)) + {}^V((L_{\varphi(V)}\varphi) - \varphi(L_V\varphi)\circ\xi) \\
&= {}^C(\varphi\circ\varphi)({}^cV) - \gamma N_V + {}^V(N_V\circ\xi) = {}^C(\varphi\circ\varphi)({}^CV) - (\gamma N)({}^CV) + {}^V(N\circ\xi)({}^CV) \\
&= ({}^C(\varphi^2) - \gamma N + {}^V(N\circ\xi))({}^CV),
\end{aligned} \tag{3.24}$$

where

$$N_V = L_{\varphi(V)}\varphi - \varphi \circ (L_V\varphi)$$

and

$$\begin{aligned}
(\Phi_\varphi\varphi)(V, W) &= (L_{\varphi(V)}\varphi - \varphi\circ(L_V\varphi))W \\
&= [\varphi V, \varphi W] - \varphi[V, \varphi W] - \varphi[\varphi V, W] + \varphi^2[V, W] \\
&= N_\varphi(V, W)
\end{aligned}$$

is nothing but the Tachibana operator or the Nijenhuis tensor $N_\varphi(V, W) \in \Im_2^1(M)$ constructed from φ.

Similarly, if $A \in \Im_q^p(M)$, then from (3.13) we have

$$\begin{aligned}
({}^C\varphi)^2({}^VA) &= ({}^C\varphi \circ {}^C\varphi){}^VA = {}^C\varphi({}^C\varphi{}^VA) = {}^C\varphi({}^V(\varphi(A))) = {}^V(\varphi(\varphi(A))) \\
&= {}^V((\varphi\circ\varphi)(A)) = {}^C(\varphi\circ\varphi){}^VA = {}^C(\varphi^2){}^VA.
\end{aligned} \tag{3.25}$$

If we take the integrability condition of φ ($N_\varphi = 0$), then by the Remark 3.2, (3.24), (3.25) and the linearity of the complete lift, we have

$$({}^C\varphi)^2 = {}^C(\varphi^2) = {}^C(-I) = -I.$$

Let M_{2n} and N_{2m} be two manifolds with complex structures φ and ψ, respectively. A differentiable mapping $f : M \to N$ is called a *holomorphic (analytic) mapping* [46, 47] if at each point $P \in M_{2n}$

$$\mathrm{d}f_P \circ \varphi_P = \psi_{f(P)} \circ \mathrm{d}f_P. \tag{3.26}$$

As the mapping $f : M_{2n} \to N_{2m}(2m = 2n + (2n)^{p+q})$ we take a pure cross-section $\sigma_\xi^\varphi : M \to T_q^p(M)$ determined by the pure tensor field $\xi \in \Im_q^p(M)$ with respect to the φ-structure. The pure cross-section $\sigma_\xi^\varphi : M \to T_q^p(M)$ can be locally expressed by (3.7). In (3.26), if ψ is the complex structure ${}^C\varphi$ (see Theorem 3.2), then the condition that the pure cross-section σ_ξ^φ be a holomorphic cross-section is locally given by

$$\varphi_l^m\, \partial_m x^K = {}^C\varphi_M^K\, \partial_l x^M, \tag{3.27}$$

where ${}^C\varphi_M^K$ are components of ${}^C\varphi$ along the pure cross-section $\sigma_\xi^\varphi(M_n)$ with respect to the natural frame $\{\partial_k, \partial_{\overline{k}}\}$.

In the case $K = k$, by virtue of (3.7) and (3.14) we get the identity $\varphi_l^k = \varphi_l^k$. When $K = \overline{k}$, by virtue of (3.7) and (3.14), (3.27) reduces to

$$(\Phi_\varphi\xi)_{lk_1\ldots k_q}^{h_1\ldots h_p} = \varphi_l^m \partial_m \xi_{k_1\ldots k_q}^{h_1\ldots h_p} - \partial_l \overset{*}{\xi}{}_{k_1\ldots k_q}^{h_1\ldots h_p} + \sum_{a=1}^{q} (\partial_{k_a}\varphi_l^m)\xi_{k_1\ldots m\ldots k_q}^{h_1\ldots h_p}$$
$$+ 2\sum_{\lambda=1}^{p} \partial_{[l}\,\varphi_{m]}^{h_\lambda} \xi_{k_1\ldots k_q}^{h_1\ldots m\ldots h_p} = 0, \tag{3.28}$$

which is the condition to be holomorphic of tensor field ξ (see Theorem 1.9), where $\Phi_\varphi\xi$ is the Tachibana operator. From (3.27) and (3.28) we have

Theorem 3.3 *Let (M_n, φ) be a complex manifold. Then the complete lift ${}^C\varphi$ of φ to the tensor bundle $T_q^p(M)$ along the pure cross-section $\sigma_\xi^\varphi(M_n)$ leaves the submanifold $\sigma_\xi^\varphi(M_n) \subset T_q^p(M)$ invariant if and only if the tensor field ξ is holomorphic with respect to.*

3.4 Dual-Holomorphic Functions and Tangent Bundles of Order 1

We consider a two-dimensional dual algebra $\mathbb{R}(\varepsilon)$, $\varepsilon^2 = 0$ (ε is nilpotent) with a standard basis $\{e_1, e_2\} = \{1, \varepsilon\}$ and structural constants $C_{\alpha\beta}^\gamma$: $e_\alpha e_\beta = C_{\alpha\beta}^\gamma e_\gamma$, $\alpha, \beta, \gamma = 1, 2,$

where $C_{11}^1 = C_{12}^2 = C_{21}^2 = 1$, $C_{12}^1 = C_{21}^1 = C_{22}^1 = C_{11}^2 = C_{22}^2 = 0$ are the components of the (1,2)-tensor $C : \mathbb{R}(\varepsilon) \times \mathbb{R}(\varepsilon) \to \mathbb{R}(\varepsilon)$.

Let $Z = x^\alpha e_\alpha$ be a variable in $\mathbb{R}(\varepsilon)$, where x^α ($\alpha = 1, 2$) are real variables. Using a real-valued C^∞-function $f^\beta(x) = f^\beta(x^1, x^2)$, $\beta = 1, 2$, we introduce a dual function $F = f^\beta(x) e_\beta$ of variable $Z \in \mathbb{R}(\varepsilon)$. It is well known that the dual function $F = F(Z)$ is holomorphic if and only if the following Scheffers condition holds (see Sect. 1.2):

$$C_2 D = D C_2, \tag{3.29}$$

where $D = \left(\frac{\partial f^\alpha}{\partial x^\beta}\right)$ is the Jacobian matrix of $f^\alpha(x)$, $C_2 = (C_{2\beta}^\gamma) = \begin{pmatrix} 0 & 0 \\ 1 & 0 \end{pmatrix}$, γ and β denotes the row and column numbers of matrix C_2, respectively. The condition (3.29) reduces to the following equations:

$$\frac{\partial f^1}{\partial x^2} = 0, \quad \frac{\partial f^2}{\partial x^2} = \frac{\partial f^1}{\partial x^1}.$$

From here it follows that the dual-holomorphic function $F = F(Z)$ has the following explicit form:

$$F(Z) = f(x^1) + \varepsilon(x^2 f'(x^1) + g(x^1)),$$

where $f(x^1) = f^1(x^1)$, $f'(x^1) = \frac{\mathrm{d}f}{\mathrm{d}x^1}$ and $g = g(x^1)$ is any real C^∞-function.

By similar devices, we see that the dual-holomorphic multi-variable function $F = F(Z^1, \ldots, Z^n)$, $Z^i = x^i + \varepsilon x^{n+i}$, $i = 1, \ldots, n$ has the form:

$$F(Z^1, \ldots, Z^n) = f(x^1, \ldots, x^n) + \varepsilon(x^{n+s} \partial_s f + g(x^1, \ldots, x^n)), \tag{3.30}$$

where $g = g(x^1, \ldots, x^n)$ is any real multi-variable C^∞-function, $\partial_s f = \frac{\partial f}{\partial x^s}$.

A *dual-holomorphic manifold* [81] $X_n(\mathbb{R}(\varepsilon))$ of dimension n is a Hausdorff space with a fixed atlas compatible with a group of $\mathbb{R}(\varepsilon)$-holomorphic transformations of space $\mathbb{R}^n(\varepsilon)$, where $\mathbb{R}^n(\varepsilon) = \mathbb{R}(\varepsilon) \times \cdots \times \mathbb{R}(\varepsilon)$ is the space of n-tupes of dual numbers $(z^1, z^2, \ldots, z^n)$ with $z^i = x^i + \varepsilon y^i \in \mathbb{R}(\varepsilon)$, $x^i, y^i \in \mathbb{R}$, $i = 1, \ldots, n$. We shall identify $\mathbb{R}^n(\varepsilon)$ with $\mathbb{R}^{2n}$, when necessary, by mapping $(z^1, z^2, \ldots, z^n) \in \mathbb{R}^n(\varepsilon)$ into $(x^1, \ldots, x^n, y^1, \ldots, y^n) \in \mathbb{R}^{2n}$ and therefore the $\mathbb{R}(\varepsilon)$-holomorphic manifold $X_n(\mathbb{R}(\varepsilon))$ is a real manifold M_{2n} of dimension $2n$.

Let now M_n be a differentiable manifold, $T(M_n)$ its tangent bundle, and π the projection $T(M_n) \to M_n$. The tangent bundle $T(M_n)$ consists of the pairs (x, v), where $x \in M_n$ and $v \in T_x(M_n)$ ($T_x(M_n)$ is the tangent vector space at $x \in M_n$). Let $(U, x = (x^1, \ldots, x^n))$ be a coordinate chart in M_n. Then it induces the local coordinates $(x^1, \ldots, x^n, x^{n+1}, \ldots, x^{2n})$ in $\pi^{-1}(U)$, where $x^{n+1}, \ldots, x^{2n}$ represent the components of $v \in T_x(M_n)$ with respect to the local frame $\{\partial_i\}$. In the following we use the notation $\bar{i} = i + n$ for all $i = 1, \ldots, n$.

If $(U', x' = (x^{1'}, \dots, x^{n'}))$ is another coordinate chart in M_n, then the induced coordinates $(x^{1'}, \dots, x^{n'}, x^{\overline{1}'}, \dots, x^{\overline{n}'})$ in $\pi^{-1}(U')$, will be given by

$$\begin{cases} x^{i'} = x^{i'}(x^i),\ i = 1, \dots, n, \\ x^{\overline{i}'} = \dfrac{\partial x^{i'}}{\partial x^i} x^{\overline{i}},\ \overline{i} = n+1, \dots, 2n. \end{cases} \tag{3.31}$$

The Jacobian of (3.31) is the matrix

$$S = \left(\frac{\partial x^{\alpha'}}{\partial x^{\alpha}}\right) = \begin{pmatrix} \dfrac{\partial x^{i'}}{\partial x^i} & 0 \\ x^{\overline{s}} \dfrac{\partial^2 x^{i'}}{\partial x^i \partial x^s} & \dfrac{\partial x^{i'}}{\partial x^i} \end{pmatrix},\ \alpha = 1, \dots, 2n\,.$$

From here follows that there exists a tensor field of type (1,1)

$$\varphi = \left(\varphi^{\alpha}_{\beta}\right) = \begin{pmatrix} \varphi^i_j & \varphi^i_{\overline{j}} \\ \varphi^{\overline{i}}_j & \varphi^{\overline{i}}_{\overline{j}} \end{pmatrix} = \begin{pmatrix} 0 & 0 \\ I & 0 \end{pmatrix} \quad (I = (\delta^i_j) \text{ is an identity matrix of degree } n) \tag{3.32}$$

with properties $\varphi^2 = 0$ and $S\varphi = \varphi S$, i.e. the transformation $S : \{\partial_\alpha\} \to \{\partial_{\alpha'}\}$ preserving φ is an admissible dual transformation. Thus, the tangent bundle $T(M_n)$ of a manifold M_n carries a natural dual structure φ, which is integrable $(\partial_k \varphi^i_j = 0)$. Therefore, with each induced coordinates $(x^i, x^{\overline{i}})$ in $\pi^{-1}(U) \subset T(M_n)$, we associate the local dual coordinates $X^i = x^i + \varepsilon x^{\overline{i}}$, $\varepsilon^2 = 0$. Using (3.31) we see that the local dual coordinates $X^i = x^i + \varepsilon x^{\overline{i}}$ are transformed by

$$X^{i'} = x^{i'}(x^i) + \varepsilon x^{\overline{s}} \partial_s (x^{i'}(x^i)). \tag{3.33}$$

Equation (3.33) shows that the quantities $X^{i'}$ are the dual-holomorphic functions of $X^i = x^i + \varepsilon x^{\overline{i}}$(see (3.30) with $g(x^1, \dots, x^n) = 0$). Thus the tangent bundle $T(M_n)$ with a natural integrable φ-structure is a real image of the dual-holomorphic manifold $X_n(\mathbb{R}(\varepsilon))$, $\dim X_n(\mathbb{R}(\varepsilon)) = n$ [81]. In such interpretation there exists a one-to-one correspondence between dual tensor fields on $X_n(\mathbb{R}(\varepsilon))$ and pure tensor fields with respect to the φ-structure on $T(M_n)$ (see Sect. 1.5).

It is important that the dual tensor field on $X_n(\mathbb{R}(\varepsilon))$ corresponding to a pure C^∞-tensor field is not necessarily dual-holomorphic. This tensor field is dual-holomorphic on $X_n(\mathbb{R}(\varepsilon))$ if and only if the Φ-operator associated with φ and applied to a pure tensor field t of type $(1, q)$ or ω of type $(0, q)$ satisfies the following conditions (see Sect. 1.6)

$$\begin{aligned}(\Phi_\varphi t)(Y, X_1, \dots, X_q) = &-(L_{t(X_1, X_2, \dots, X_q)}\varphi)Y \\ &+ \sum_{\lambda=1}^{q} t(X_1, X_2, \dots, (L_{X_\lambda}\varphi)Y, \dots, X_q) = 0\end{aligned}$$

or

$$(\Phi_\varphi \omega)(Y, X_1, \ldots, X_q) = (\varphi Y)(\omega(X_1, X_2, \ldots, X_q)) - Y(\omega(\varphi X_1, X_2, \ldots, X_q)) + \sum_{\lambda=1}^{q} \omega(X_1, X_2, \ldots, \varphi(L_Y X_\lambda), \ldots, X_q) = 0,$$

where L_Y is the Lie derivation with respect to Y.

From (3.30) we immediately have

$$F = {}^V f + \varepsilon({}^C f + {}^V g),$$

where g is any function on M_n, ${}^V f = f \circ \pi$, ${}^V g = g \circ \pi$ are the vertical lifts of f, g and ${}^C f = x^{n+s} \partial_s f$ is the complete lift of f from M_n to its tangent bundle $T(M_n)$ (see [87]). The study the theory of lifts in the tangent bundles was started by Yano and Kobayashi [85] (see also [69]) and devoleped by many authors (see for example [1, 5, 8, 14, 15, 24, 42, 43, 58, 59, 78]). We call ${}^D f = {}^C f + {}^V g$ the *deformed complete lift of a function* f to thc tangent bundle $T(M_n)$.

Thus we have

Theorem 3.4 *Let $T(M_n)$ be a tangent bundle of M_n, which is a real image of the dual-holomorphic manifold $X_n(\mathbb{R}(\varepsilon))$. Then the vertical and the deformed complete lifts to $T(M_n)$ of any function on M_n are the real and dual part of corresponding dual-holomorphic function on $X_n(\mathbb{R}(\varepsilon))$, respectively.*

3.5 Deformed Complete Lifts of Vector Fields

In a tangent bundle $T(M_n)$ with dual structure φ, a vector field $\tilde{V} = (\tilde{v}^\alpha) = (\tilde{v}^i, \tilde{v}^{n+i}) = (\tilde{v}^i, \tilde{v}^{\bar{i}})$ is called a *dual-holomorphic vector field* if $L_{\tilde{V}}\varphi = 0$. Such a vector field is the real image of corresponding dual-holomorphic vector field $V = (V^i)$ on $X_n(\mathbb{R}(\varepsilon))$, where $V^i = \tilde{v}^i + \tilde{v}^{\bar{i}}\varepsilon$. The condition of dual-holomorphy of a vector field $\tilde{V}$ on $T(M_n)$ may be now locally written as follows:

$$L_{\tilde{V}}\varphi_\beta^\alpha = \tilde{v}^\sigma \partial_\sigma \varphi_\beta^\alpha - (\partial_\sigma \tilde{v}^\alpha)\varphi_\beta^\sigma + (\partial_\beta \tilde{v}^\sigma)\varphi_\sigma^\alpha = 0\,. \tag{3.34}$$

By virtue of (3.32), we have

(a) The case where $\alpha = i$, $\beta = j$, the Eq. (3.34) reduces to

$$\begin{aligned} L_{\tilde{V}}\varphi_j^i &= \tilde{v}^\sigma \partial_\sigma \varphi_j^i - (\partial_\sigma \tilde{v}^i)\varphi_j^\sigma + (\partial_j \tilde{v}^\sigma)\varphi_\sigma^i \\ &= \tilde{v}^m \partial_m \varphi_j^i + \tilde{v}^{\overline{m}} \partial_{\overline{m}} \varphi_j^i - (\partial_m \tilde{v}^i)\varphi_j^m - (\partial_{\overline{m}} \tilde{v}^i)\varphi_j^{\overline{m}} + (\partial_j \tilde{v}^m)\varphi_m^i + (\partial_j \tilde{v}^{\overline{m}})\varphi_{\overline{m}}^i \end{aligned}$$

$$= -(\partial_{\overline{m}}\tilde{v}^i)\delta_j^m = -(\partial_{\overline{j}}\tilde{v}^i) = 0,$$

from which it follows

$$\tilde{v}^i = v^i(x^1, \ldots, x^n). \tag{3.35}$$

(b) In the case where $\alpha = i,\ \beta = \overline{j}$ and $\alpha = \overline{i},\ \beta = \overline{j}$, the Eq. (3.34) reduces to $0 = 0$.
(c) In the case where $\alpha = \overline{i},\ \beta = j$, the Eq. (3.34) reduces to

$$\begin{aligned}
L_{\tilde{V}}\varphi_j^{\overline{i}} &= \tilde{v}^\sigma \partial_\sigma \varphi_j^{\overline{i}} - (\partial_\sigma \tilde{v}^{\overline{i}})\varphi_j^\sigma + (\partial_j \tilde{v}^\sigma)\varphi_\sigma^{\overline{i}} \\
&= \tilde{v}^m \partial_m \varphi_j^{\overline{i}} + \tilde{v}^{\overline{m}} \partial_{\overline{m}} \varphi_j^{\overline{i}} - (\partial_m \tilde{v}^{\overline{i}})\varphi_j^m - (\partial_{\overline{m}} \tilde{v}^{\overline{i}})\varphi_j^{\overline{m}} + (\partial_j \tilde{v}^m)\varphi_m^{\overline{i}} + (\partial_j \tilde{v}^{\overline{m}})\varphi_{\overline{m}}^{\overline{i}} \\
&= -(\partial_{\overline{m}} \tilde{v}^{\overline{i}})\varphi_j^{\overline{m}} + (\partial_j \tilde{v}^m)\varphi_m^{\overline{i}} = -(\partial_{\overline{m}} \tilde{v}^{\overline{i}})\delta_j^m + (\partial_j \tilde{v}^m)\delta_m^i = 0,
\end{aligned}$$

from which it follows

$$\partial_{\overline{j}}\tilde{v}^{\overline{i}} = \partial_j v^i,$$

and after integrating, we find

$$\tilde{v}^{\overline{i}} = x^{\overline{j}}\partial_j v^i + w^i(x^1, \ldots, x^n), \tag{3.36}$$

where $w^i = w^i(x^1, \ldots, x^n)$ are any real multi-variable C^∞-functions.

Remark 3.4 Using (3.31), (3.35), (3.36) and $\tilde{v}^{\alpha'} = \frac{\partial x^{\alpha'}}{\partial x^\alpha}\tilde{v}^\alpha$, we easily see that $v = (v^i(x^1, \ldots, x^n))$ and $w = (w^i(x^1, \ldots, x^n))$ are vector fields on M_n.

Thus a real dual-holomorphic vector field $\tilde{V}$ on the tangent bundle can be written in the form

$$\tilde{V} = (\tilde{v}^\alpha) = \begin{pmatrix} \tilde{v}^i \\ \tilde{v}^{\overline{i}} \end{pmatrix} = \begin{pmatrix} v^i(x^1, \ldots, x^n) \\ x^{\overline{j}}\partial_j v^i + w^i(x^1, \ldots, x^n) \end{pmatrix} = \begin{pmatrix} v^i \\ x^{\overline{j}}\partial_j v^i \end{pmatrix} + \begin{pmatrix} 0 \\ w^i \end{pmatrix} = {}^C v + {}^V w,$$

where ${}^C v$ and ${}^V w$ are the complete and vertical lifts of vector fields $v = (v^i)$ and $w = (w^i)$ from M_n to the tangent bundle $T(M_n)$, respectively [87]. Thus we have

Theorem 3.5 *Let $T(M_n)$ be the tangent bundle of M_n, which is the real image of dual-holomorphic manifold $X_n(\mathbb{R}(\varepsilon))$. Then the real image of corresponding dual-holomorphic*

vector field $V = (V^i) = (\tilde{v}^i + \tilde{v}^{\bar{i}}\varepsilon)$ *is a deformed complete lift in the form* ${}^D V = {}^C v + {}^V w$, *where* ${}^C v$ *and* ${}^V w$ *are the complete and vertical lifts of the vector fields* $v = (v^i)$ *and* $w = (w^i)$ *from* M_n *to* $T(M_n)$, *respectively.*

3.6 Deformed Complete Lifts of Tensor Fields of Type (1,1)

A tensor field $\tilde{t}$ of type (1,1) on the tangent bundle $T(M_n)$ is called *pure* with respect to the dual structure φ if

$$\tilde{t}(\varphi X) = \varphi(\tilde{t}X),$$

for any vector field X on $T(M_n)$. From here we see that the condition of pure tensor fields may be expressed in terms of the local induced coordinates as follows:

$$\tilde{t}_\sigma^\beta \varphi_\alpha^\sigma = \tilde{t}_\alpha^\sigma \varphi_\sigma^\beta.$$

Using (3.32), from the last condition we have

$$\tilde{t} = \left(\tilde{t}_\beta^\alpha\right) = \begin{pmatrix} \tilde{t}_j^i & 0 \\ \tilde{t}_j^{\bar{i}} & \tilde{t}_j^i \end{pmatrix}. \tag{3.37}$$

A pure tensor field $\tilde{t}$ is called a *dual-holomorphic tensor field* if $\Phi_\varphi t = 0$, where Φ_φ is the Tachibana operator defined by

$$(\Phi_\varphi \tilde{t})(X, Y) = [\varphi X, \tilde{t}Y] - \varphi[X, \tilde{t}Y] - \tilde{t}[\varphi X, Y] + \varphi\tilde{t}[X, Y].$$

We note that, such a tensor field is the real image of corresponding dual-holomorphic tensor field from $X_n(\mathbb{R}(\varepsilon))$. Sometimes the tensor $\Phi_\varphi \tilde{t}$ of type (1,2) is called the Nijenhuis-Shirokov tensor field. It is clear that, if $\varphi = \tilde{t}$, then $\Phi_\varphi \tilde{t}$ is the Nijenhuis tensor field N_φ, i.e. $\Phi_\varphi \varphi = N_\varphi$.

The condition of dual-holomorphy of a pure tensor field $\tilde{t}$ on $T(M_n)$ may be now locally written as follows:

$$(\Phi_\varphi \tilde{t})_{\gamma\beta}^\alpha = \varphi_\gamma^\sigma \partial_\sigma \tilde{t}_\beta^\alpha - \varphi_\sigma^\alpha \partial_\gamma \tilde{t}_\beta^\sigma - \tilde{t}_\beta^\sigma \partial_\sigma \varphi_\gamma^\alpha + \tilde{t}_\sigma^\alpha \partial_\beta \varphi_\gamma^\sigma = 0. \tag{3.38}$$

By virtue of (3.32) and (3.37), the Eq. (3.38) after some calculations reduces to

$$\partial_{\bar{k}} \tilde{t}_j^i = 0, \quad \partial_{\bar{k}} \tilde{t}_j^{\bar{i}} - \partial_k \tilde{t}_j^i = 0.$$

From here it follows that

$$\tilde{t}_j^i = t_j^i(x^1, \ldots, x^n), \quad \tilde{t}_j^{\bar{i}} = x^{\bar{k}} \partial_k t_j^i + g_j^i, \tag{3.39}$$

where $g^i_j = g^i_j(x^1, \ldots, x^n)$.

Remark 3.5 Using (3.31), (3.39) and $t^{\alpha'}_{\beta'} = \frac{\partial x^{\alpha'}}{\partial x^{\alpha}} \frac{\partial x^{\beta}}{\partial x^{\beta'}} t^{\alpha}_{\beta}$, we easily see that $t^i_j(x^1, \ldots, x^n)$ and $g^i_j\ (x^1, \ldots, x^n)$ are components of any tensor fields t and g of type (1,1) on M_n.

Thus, a dual-holomorphic tensor field $\tilde{t}$ on the tangent bundle can be written in the form

$$\tilde{t} = \left(\tilde{t}^{\alpha}_{\beta}\right) = \begin{pmatrix} t^i_j & 0 \\ x^{\bar{k}} \partial_k t^i_j + g^i_j & t^i_j \end{pmatrix} = \begin{pmatrix} t^i_j & 0 \\ x^{\bar{k}} \partial_k t^i_j & t^i_j \end{pmatrix} + \begin{pmatrix} 0 & 0 \\ g^i_j & 0 \end{pmatrix} = {}^C t + {}^V g,$$

where ${}^C t$ and ${}^V g$ are the complete and vertical lifts of (1,1)-tensor fields t and g from M_n to tangent bundle $T(M_n)$, respectively (see [87]). Thus we have

Theorem 3.6 *Let $T(M_n)$ be the tangent bundle of M_n, which is the real image of the dual-holomorphic manifold $X_n(\mathbb{R}(\varepsilon))$. Then the real image of corresponding dual-holomorphic tensor field T of type (1,1) from $X_n(\mathbb{R}(\varepsilon))$ is the deformed complete lift in the form ${}^D t = {}^C t + {}^V g$, where ${}^C t$ and ${}^V g$ are the complete and vertical lifts of the (1,1)-tensor fields t and g from M_n to $T(M_n)$, respectively.*

Let (M_{4n}, F, G, H) be an almost quaternion manifold, i.e.

$$\begin{aligned} &F^2 = -I, \quad G^2 = -I, \quad H^2 = -I, \\ &F = GH = -HG, \quad G = HF = -FH, \quad H = FG = -GF. \end{aligned}$$

Then for three tensor fields F, G and H of type (1,1), we now consider the deformed complete lifts:

$${}^D F = {}^C F + {}^V G, \quad {}^D G = {}^C G + {}^V H, \quad {}^D H = {}^C H + {}^V F.$$

From here, we find

$$\begin{aligned} ({}^D F)^2 &= \begin{pmatrix} F^i_m & 0 \\ x^{\bar{s}} \partial_s F^i_m + G^i_m & F^i_m \end{pmatrix} \begin{pmatrix} F^m_j & 0 \\ x^{\bar{s}} \partial_s F^m_j + G^m_j & F^m_j \end{pmatrix} \\ &= \begin{pmatrix} F^i_m F^m_j & 0 \\ x^{\bar{s}} (\partial_s F^i_m) F^m_j + G^i_m F^m_j + F^i_m x^{\bar{s}} \partial_s F^m_j + F^i_m G^m_j & F^i_m F^m_j \end{pmatrix} \\ &= \begin{pmatrix} F^2 & 0 \\ x^{\bar{s}} \partial_s F^2 + GF + FG & F^2 \end{pmatrix} = \begin{pmatrix} -I_{M_n} & 0 \\ 0 & -I_{M_n} \end{pmatrix} = -I_{T(M_n)}. \end{aligned}$$

Similarly we get

$$({}^D G)^2 = -I_{T(M_n)}, \quad ({}^D H)^2 = -I_{T(M_n)}.$$

Thus we have

Theorem 3.7 *Let (M_{4n}, F, G, H) be an almost quaternion manifold. Then the deformed complete lift of each structure F, G and H is an almost complex structure on the tangent bundle $T(M_{4n})$.*

3.7 Deformed Complete Lifts of 1-Forms

A 1-form $\tilde{\omega}$ on the tangent bundle $T(M_n)$ is called a dual-holomorphic 1-form, if $\Phi_\varphi \tilde{\omega} = 0$, where Φ_φ is the Tachibana operator defined by

$$(\Phi_\varphi \tilde{\omega})(X, Y) = (\varphi X)(\tilde{\omega}(Y)) - X(\tilde{\omega}(\varphi Y) + \tilde{\omega}((L_Y \varphi)X).$$

Such a 1-form is a real image of corresponding dual-holomorphic 1-form from $X_n(\mathbb{R}(\varepsilon))$. The tensor field $\Phi_\varphi \tilde{\omega}$ of type (0,2) has components

$$(\Phi_\varphi \tilde{\omega})_{\alpha\beta} = \varphi_\alpha^\sigma \partial_\sigma \tilde{\omega}_\beta - \varphi_\beta^\sigma \partial_\alpha \tilde{\omega}_\sigma - \tilde{\omega}_\sigma (\partial_\alpha \varphi_\beta^\sigma - \partial_\beta \varphi_\alpha^\sigma)$$

with respect to the natural frame $\{\partial_\alpha\} = \{\partial_i, \partial_{\bar{i}}\}$.

By virtue of (3.32), the equation $(\Phi_\varphi \tilde{\omega})_{\alpha\beta} = 0$ reduces to

$$\partial_{\bar{i}} \tilde{\omega}_j - \partial_i \tilde{\omega}_{\bar{j}} = 0, \quad \partial_{\bar{i}} \tilde{\omega}_{\bar{j}} = 0.$$

From here we have

$$\tilde{\omega}_{\bar{j}} = \omega_j(x^1, \ldots, x^n), \quad \tilde{\omega}_j = x^{\bar{i}} \partial_i \omega_j + \theta_j(x^1, \ldots, x^n). \tag{3.40}$$

Remark 3.6 Using (3.31), (3.40) and $\tilde{\omega}_{\beta'} = \frac{\partial x^\beta}{\partial x^{\beta'}} \tilde{\omega}_\beta$, we easily see that $\omega_j(x^1, \ldots, x^n)$ and $\theta_j\,(x^1, \ldots, x^n)$ are the components of any 1-forms ω and θ on M_n, respectively.

Thus, a real dual-holomorphic 1-form $\tilde{\omega}$ on tangent bundle can be rewritten in the form

$$\tilde{\omega} = (\tilde{\omega}_j, \tilde{\omega}_{\bar{j}}) = (x^{\bar{i}} \partial_i \omega_j + \theta_j, \omega_j) = (x^{\bar{i}} \partial_i \omega_j, \omega_j) + (\theta_j, 0) = {}^C\omega + {}^V\theta,$$

where ${}^C\omega$ and ${}^V\theta$ are the complete and vertical lifts of 1-forms $\omega = (\omega_j)$ and $\theta = (\theta_j)$ from M_n to its tangent bundle $T(M_n)$, respectively (see [87]). Thus we have

Theorem 3.8 *Let $T(M_n)$ be the tangent bundle of M_n, which is the real image of the dual-holomorphic manifold $X_n(\mathbb{R}(\varepsilon))$. Then the real image of the corresponding dual-holomorphic 1-form from $X_n(\mathbb{R}(\varepsilon))$ is a deformed complete lift in the form ${}^D\omega = {}^C\omega + {}^V\theta$, where ${}^C\omega$ and ${}^V\theta$ are the complete and vertical lifts of 1-forms $\omega = (\omega_j)$ and $\theta = (\theta_j)$ from M_n to $T(M_n)$, respectively.*

3.8 Deformed Complete Lifts of Riemannian Metrics

A tensor field $\tilde{g}$ of type (0,2) on the tangent bundle $T(M_n)$ is called a pure tensor field with respect to the dual structure φ if

$$\tilde{g}(\varphi X, Y) = \tilde{g}(X, \varphi Y),$$

for any vector fields X and Y on $T(M_n)$. From here we see that the condition of purity of $\tilde{g}$ may be expressed in terms of the local induced coordinates as follows:

$$\tilde{g}_{\sigma\beta}\varphi^{\sigma}_{\alpha} = \tilde{g}_{\alpha\sigma}\varphi^{\sigma}_{\beta}.$$

Using (3.32), from the last condition we have

$$\tilde{g} = (\tilde{g}_{\alpha\beta}) = \begin{pmatrix} \tilde{g}_{ij} & \tilde{g}_{\bar{i}j} \\ \tilde{g}_{\bar{i}j} & 0 \end{pmatrix}, \quad \tilde{g}_{\bar{i}\bar{j}} = 0, \quad \tilde{g}_{\bar{i}j} = \tilde{g}_{i\bar{j}}.$$

A pure tensor field $\tilde{g}$ of type (0,2) on the tangent bundle $T(M_n)$ is called dual-holomorphic with respect to φ, if $\Phi_{\varphi}\tilde{g} = 0$, where Φ_{φ} is the Tachibana operator defined by

$$\begin{aligned}(\Phi_{\varphi}\tilde{g})(X, Y, Z) &= (\varphi X)(\tilde{g}(Y, Z)) - X(\tilde{g}(\varphi Y, Z)) \\ &\quad + \tilde{g}((L_Y\varphi)X, Z) + \tilde{g}(Y, (L_Z\varphi)X),\end{aligned}$$

for every $X, Y, Z \in \mathfrak{I}^1_0(T(M_n))$. Such a tensor field is the real image of the corresponding dual-holomorphic tensor field of type (0,2) from $X_n(\mathbb{R}(\varepsilon))$. It is well known that, if $\tilde{g}$ is a Riemannian metric and $\nabla^{\tilde{g}}$ its Levi–Civita connection, then the condition $\Phi_{\varphi}\tilde{g} = 0$ is equivalent to the condition $\nabla^{\tilde{g}}\varphi = 0$ [25], i.e. the triple $(T(M_n), \tilde{g}, \varphi)$ is a dual anti-Kähler (or Kähler-Norden) manifold.

The tensor field $\Phi_{\varphi}\tilde{g}$ of type (0,3) has the components

$$(\Phi_{\varphi}\tilde{g})_{\alpha\beta\gamma} = \varphi^{\sigma}_{\alpha}\partial_{\sigma}\tilde{g}_{\beta\gamma} - \varphi^{\sigma}_{\beta}\partial_{\alpha}\tilde{g}_{\sigma\gamma} - \tilde{g}_{\sigma\gamma}(\partial_{\alpha}\varphi^{\sigma}_{\beta} - \partial_{\beta}\varphi^{\sigma}_{\alpha}) + \tilde{g}_{\beta\sigma}\partial_{\gamma}\varphi^{\sigma}_{\alpha}$$

with respect to the natural frame $\{\partial_{\alpha}\} = \{\partial_i, \partial_{\bar{i}}\}$.

By virtue of (3.32), after some calculations, the equation $(\Phi_{\varphi}\tilde{g})_{\alpha\beta\gamma} = 0$ reduces to

$$\partial_{\bar{i}}\tilde{g}_{jk} - \partial_i\tilde{g}_{\bar{j}k} = 0, \quad \partial_{\bar{i}}\tilde{g}_{\bar{j}k} = 0,$$

from which we have

$$\tilde{g}_{\bar{j}k} = g_{jk}(x^1, \ldots, x^n), \quad \tilde{g}_{jk} = x^{\bar{i}}\partial_i g_{jk} + h_{jk}(x^1, \ldots, x^n). \tag{3.41}$$

Remark 3.7 Using (3.31), (3.41) and $\tilde{g}_{\alpha'\beta'} = \frac{\partial x^\alpha}{\partial x^{\alpha'}}\frac{\partial x^\beta}{\partial x^{\beta'}}\tilde{g}_{\alpha\beta}$, we easily see that $g_{jk}(x^1, \ldots, x^n)$ and $h_{jk}(x^1, \ldots, x^n)$ are components of any tensor fields g and h of type (0,2) on M_n, respectively.

Thus a real dual-holomorphic tensor field $\tilde{g}$ of type (0,2) on tangent bundle can be rewritten in the form

$$\tilde{g} = (\tilde{g}_{\beta\gamma}) = \begin{pmatrix} x^{\bar{i}}\partial_i g_{jk} + h_{jk} & g_{jk} \\ g_{jk} & 0 \end{pmatrix}$$
$$= \begin{pmatrix} x^{\bar{i}}\partial_i g_{jk} & g_{jk} \\ g_{jk} & 0 \end{pmatrix} + \begin{pmatrix} h_{jk} & 0 \\ 0 & 0 \end{pmatrix} = {}^C g + {}^V h,$$

where ${}^C g$ and ${}^V h$ are the complete and vertical lifts of the tensor fields $g = (g_{jk})$ and $h = (h_{jk})$ of type (0,2) from M_n to the tangent bundle $T(M_n)$, respectively (see [87]). Therefore we have

Theorem 3.9 *Let $T(M_n)$ be the tangent bundle of M_n, which is the real image of the dual-holomorphic manifold $X_n(\mathbb{R}(\varepsilon))$. Then the real image of corresponding dual-holomorphic tensor field of type (0,2) from $X_n(\mathbb{R}(\varepsilon))$ is a deformed complete lift in the form ${}^D g = {}^C g + {}^V h$, where ${}^C g$ and ${}^V h$ are the complete and vertical lifts of $g = (g_{jk})$ and $h = (h_{jk})$ from M_n to $T(M_n)$, respectively.*

Remark 3.8 Let now g be a Riemannian metric, and h be any symmetric (0,2)-tensor field on M_n. It is clear that in such case the tensor field ${}^D g = {}^C g + {}^V h$ is a Riemannian metric on $T(M_n)$. We note that lifts of this kind have been also studied under the names: the metric I + II [87] if $g = h$ and the synectic lift [78].

3.9 Deformed Complete Lifts of Connections

Let $\tilde{\nabla}$ be a connection with the components $\tilde{\Gamma}^\gamma_{\alpha\beta}$ on the tangent bundle $T(M_n)$ preserving the structure φ. That connection is called a pure connection by definition if

$$\tilde{\Gamma}^\sigma_{\alpha\beta}\varphi^\gamma_\sigma = \tilde{\Gamma}^\gamma_{\sigma\beta}\varphi^\sigma_\alpha = \tilde{\Gamma}^\gamma_{\alpha\sigma}\varphi^\sigma_\beta.$$

Using (3.32), from the purity condition we have

$$\tilde{\Gamma}^{k}_{\bar{i}j} = \tilde{\Gamma}^{k}_{i\bar{j}} = \tilde{\Gamma}^{k}_{\bar{i}\bar{j}} = \tilde{\Gamma}^{\bar{k}}_{\bar{i}\bar{j}} = 0. \tag{3.42}$$

The pure connection $\tilde{\nabla}$ with components $\tilde{\Gamma}^{\gamma}_{\alpha\beta}$ is called a dual-holomorphic connection, if [60]

$$(\Phi_{\varphi}\Gamma)^{\gamma}_{\tau\alpha\beta} = \varphi^{\sigma}_{\tau}\,\partial_{\sigma}\tilde{\Gamma}^{\gamma}_{\alpha\beta} - \varphi^{\sigma}_{\alpha}\,\partial_{\tau}\tilde{\Gamma}^{\gamma}_{\sigma\beta} = 0.$$

It is well known that, such a connection is a real image of corresponding dual-holomorphic connection from $X_n(\mathbb{R}(\varepsilon))$. From here, by virtue of (3.32) and (3.42), we have

$$(\Phi_{\varphi}\Gamma)^{k}_{tij} = \varphi^{m}_{t}\,\partial_{m}\tilde{\Gamma}^{k}_{ij} + \varphi^{\bar{m}}_{t}\,\partial_{m}\tilde{\Gamma}^{k}_{ij} - \varphi^{m}_{i}\,\partial_{t}\tilde{\Gamma}^{k}_{mj} - \varphi^{\bar{m}}_{i}\,\partial_{t}\tilde{\Gamma}^{k}_{mj} = 0$$
$$\Leftrightarrow \tilde{\Gamma}^{k}_{ii} = \Gamma^{k}_{ii}(x^{1}, \ldots, x^{n}),$$
$$(\Phi_{\varphi}\Gamma)^{k}_{iij} = \varphi^{m}_{t}\,\partial_{m}\tilde{\Gamma}^{k}_{ij} + \varphi^{\bar{m}}_{t}\,\partial_{\bar{m}}\tilde{\Gamma}^{k}_{ij} - \varphi^{m}_{i}\,\partial_{i}\tilde{\Gamma}^{k}_{mj} - \varphi^{\bar{m}}_{i}\,\partial_{i}\tilde{\Gamma}^{k}_{\pi j} = 0 \Leftrightarrow 0 = 0,$$
$$(\Phi_{\varphi}\Gamma)^{k}_{t\bar{i}j} = \varphi^{m}_{t}\,\partial_{m}\tilde{\Gamma}^{k}_{\bar{i}j} + \varphi^{m}_{t}\,\partial_{\bar{m}}\tilde{\Gamma}^{k}_{\bar{i}j} - \varphi^{m}_{\bar{i}}\,\partial_{t}\tilde{\Gamma}^{k}_{mj} - \varphi^{m}_{\bar{i}}\,\partial_{t}\tilde{\Gamma}^{k}_{mjj} = 0 \Leftrightarrow 0 = 0,$$
$$(\Phi_{\varphi}\Gamma)^{k}_{ti\bar{j}} = \varphi^{m}_{t}\,\partial_{mm}\tilde{\Gamma}^{k}_{i\bar{j}} + \varphi^{m}_{t}\,\partial_{\bar{m}}\tilde{\Gamma}^{k}_{i\bar{j}} - \varphi^{m}_{i}\,\partial_{t}\tilde{\Gamma}^{k}_{m\bar{j}} - \varphi^{m}_{i}\,\partial_{t}\tilde{\Gamma}^{k}_{\overline{mj}} = 0 \Leftrightarrow 0 = 0,$$
$$(\Phi_{\varphi}\Gamma)^{k}_{TTT} = \varphi^{m}_{\bar{T}}\,\partial_{m}\tilde{\Gamma}^{k}_{i\bar{J}} + \varphi^{\bar{m}}_{\bar{T}}\,\partial_{\bar{m}}\tilde{\Gamma}^{k}_{T\bar{J}} - \varphi^{m}_{t}\,\partial_{\bar{T}}\tilde{\Gamma}^{k}_{mj} - \varphi^{\bar{m}}_{i}\,\partial_{\bar{T}}\tilde{\Gamma}^{k}_{mj} = 0 \Leftrightarrow 0 = 0,$$
$$\Big(\Phi_{\varphi}\Gamma^{\bar{k}}_{tij} = \varphi^{m}_{t}\,\partial_{m}\tilde{\Gamma}^{\bar{k}}_{ij} + \varphi^{\bar{m}}_{t}\,\partial_{\bar{m}}\tilde{\Gamma}^{\bar{k}}_{ij} - \varphi^{m}_{i}\,\partial_{\bar{t}}\tilde{\Gamma}^{\bar{k}}_{mj} - \varphi^{\bar{m}}_{i}\,\partial_{i}\tilde{\Gamma}^{\bar{k}}_{mj} = 0 \Leftrightarrow \varphi^{\bar{m}}_{i}\,\partial_{\bar{t}}\tilde{\Gamma}^{\bar{k}}_{mj} = 0$$
$$\Leftrightarrow \tilde{\Gamma}^{k}_{T_i} = \Gamma^{k}_{ii}(x^{1}, \ldots, x^{n})$$
$$(\Phi_{\varphi}\Gamma)^{k}_{\bar{i}i\bar{j}} = \varphi^{m}_{T}\,\partial_{m}\tilde{\Gamma}^{k}_{i\bar{j}} + \varphi^{\bar{m}}_{T}\,\partial_{\bar{m}}\tilde{\Gamma}^{k}_{i\bar{j}} - \varphi^{m}_{i}\,\partial_{T}\tilde{\Gamma}^{k}_{m\bar{j}} - \varphi^{m}_{i}\,\partial_{T}\tilde{\Gamma}^{k}_{m\bar{j}} = 0 \Leftrightarrow 0 = 0,$$
$$(\Phi_{\varphi}\Gamma)^{k}_{\bar{t}\bar{i}j} = \varphi^{m}_{\bar{t}}\,\partial_{m}\tilde{\Gamma}^{k}_{\bar{i}j} + \varphi^{\bar{m}}_{\bar{m}}\,\partial_{\bar{m}}\tilde{\Gamma}^{k}_{\bar{i}j} - \varphi^{m}_{\bar{i}}\,\partial_{\bar{T}}\tilde{\Gamma}^{k}_{mj} - \varphi^{\bar{m}}_{\bar{i}}\,\partial_{\bar{T}}\tilde{\Gamma}^{k}_{mjj} = 0 \Leftrightarrow 0 = 0,$$
$$(\Phi_{\varphi}\Gamma)^{k}_{tij} = \varphi^{m}_{t}\,\partial_{m}\tilde{\Gamma}^{k}_{Tj} + \varphi^{m}_{t}\,\partial_{\min}\tilde{\Gamma}^{k}_{ij} - \varphi^{m}_{T}\,\partial_{t}\tilde{\Gamma}^{k}_{mjj} - \varphi^{m}_{T}\,\partial_{t}\tilde{\Gamma}^{k}_{mj} = 0 \Leftrightarrow 0 = 0,$$
$$(\Phi_{\varphi}\Gamma)^{\bar{k}}_{tij} = \varphi^{m}_{t}\,\partial_{m}\tilde{\Gamma}^{\bar{k}}_{\bar{i}j} + \varphi^{\bar{m}}_{t}\,\partial_{\bar{m}}\tilde{\Gamma}^{\bar{k}}_{ij} - \varphi^{m}_{t}\,\partial_{t}\tilde{\Gamma}^{\bar{k}}_{mj} - \varphi^{\bar{m}}_{t}\,\partial_{t}\tilde{\Gamma}^{\bar{k}}_{mj} = 0 \Leftrightarrow 0 = 0,$$
$$(\Phi_{\varphi}\Gamma)^{\bar{k}}_{tij} = \varphi^{m}_{t}\,\partial_{m}\tilde{\Gamma}^{\bar{k}}_{ij} + \varphi^{\bar{m}}_{t}\,\partial_{\bar{m}}\tilde{\Gamma}^{\bar{k}}_{i\bar{j}} - \varphi^{m}_{i}\,\partial_{t}\tilde{\Gamma}^{\bar{k}}_{m\bar{j}} - \varphi^{\bar{m}}_{i}\,\partial_{t}\tilde{\Gamma}^{\bar{k}}_{\overline{mjj}} = 0 \Leftrightarrow \varphi^{\bar{m}}_{i}\,\partial_{\bar{N}}\tilde{\Gamma}^{k}_{i\bar{T}} = 0$$
$$\Leftrightarrow \tilde{\Gamma}^{k}_{i\bar{k}} = \Gamma^{k}_{ii}(x^{1}, \ldots, x^{n}),$$
$$(\Phi_{\varphi}\Gamma)^{\bar{k}}_{tij} = \varphi^{m}_{t}\,\partial_{m}\tilde{\Gamma}^{\bar{k}}_{ij} + \varphi^{\bar{m}}_{t}\,\partial_{\bar{m}}\tilde{\Gamma}^{\bar{k}}_{ij} - \varphi^{m}_{i}\,\partial_{t}\tilde{\Gamma}^{\bar{k}}_{mj} - \varphi^{\bar{m}}_{i}\,\partial_{t}\tilde{\Gamma}^{\bar{k}}_{\bar{m}j} = 0$$
$$\Leftrightarrow \varphi^{\bar{m}}_{t}\,\partial_{\bar{m}}\tilde{\Gamma}^{\bar{k}}_{ij} - \varphi^{\bar{m}}_{i}\,\partial_{t}\tilde{\Gamma}^{\bar{k}}_{\bar{m}j} = 0 \Leftrightarrow \partial_{\bar{t}}\tilde{\Gamma}^{\bar{k}}_{ij} - \partial_{t}\tilde{\Gamma}^{\bar{k}}_{\bar{i}j} = 0$$
$$\Leftrightarrow \tilde{\Gamma}^{\bar{k}}_{ij} = x^{\bar{t}}\,\partial_{t}\Gamma^{k}_{ij} + H^{k}_{ij}(x^{1}, \ldots, x^{n}),$$
$$(\Phi_{\varphi}\Gamma)^{\bar{k}}_{\bar{t}\bar{i}j} = \varphi^{m}_{\bar{t}}\,\partial_{m}\tilde{\Gamma}^{\bar{k}}_{\bar{i}j} + \varphi^{\bar{m}}_{\bar{t}}\,\partial_{\bar{m}}\tilde{\Gamma}^{\bar{k}}_{\bar{i}j} - \varphi^{m}_{\bar{i}}\,\partial_{\bar{t}}\tilde{\Gamma}^{\bar{k}}_{mj} - \varphi^{\bar{m}}_{\bar{i}}\,\partial_{\bar{t}}\tilde{\Gamma}^{\bar{k}}_{\bar{m}j} = 0 \Leftrightarrow 0 = 0,$$
$$(\Phi_{\varphi}\Gamma)^{\bar{k}}_{t\bar{i}\bar{j}} = \varphi^{m}_{t}\,\partial_{m}\tilde{\Gamma}^{\bar{k}}_{\bar{i}\bar{j}} + \varphi^{\bar{m}}_{t}\,\partial_{\bar{m}}\tilde{\Gamma}^{\bar{k}}_{\bar{i}\bar{j}} - \varphi^{m}_{\bar{i}}\,\partial_{t}\tilde{\Gamma}^{\bar{k}}_{\bar{m}j} - \varphi^{\bar{m}}_{\bar{i}}\,\partial_{t}\tilde{\Gamma}^{\bar{k}}_{\bar{m}\bar{j}} = 0 \Leftrightarrow 0 = 0,$$

$$(\Phi_\varphi\Gamma)^{\bar{k}}_{\bar{i}i\bar{j}} = \varphi^m_{\bar{i}}\partial_m\tilde{\Gamma}^{\bar{k}}_{i\bar{j}} + \varphi^{\bar{m}}_{\bar{i}}\partial_{\bar{m}}\tilde{\Gamma}^{\bar{k}}_{i\bar{j}} - \varphi^m_i\partial_{\bar{i}}\tilde{\Gamma}^{\bar{k}}_{m\bar{j}} - \varphi^{\bar{m}}_i\partial_{\bar{i}}\tilde{\Gamma}^{\bar{k}}_{\bar{m}\bar{j}} = 0 \Leftrightarrow 0 = 0,$$
$$(\Phi_\varphi\Gamma)^{\bar{k}}_{\bar{i}\bar{i}\bar{j}} = \varphi^m_{\bar{i}}\partial_m\tilde{\Gamma}^{\bar{k}}_{\bar{i}\bar{j}} + \varphi^{\bar{m}}_{\bar{i}}\partial_{\bar{m}}\tilde{\Gamma}^{\bar{k}}_{\bar{i}\bar{j}} - \varphi^m_{\bar{i}}\partial_{\bar{i}}\tilde{\Gamma}^{\bar{k}}_{m\bar{j}} - \varphi^{\bar{m}}_{\bar{i}}\partial_{\bar{i}}\tilde{\Gamma}^{\bar{k}}_{\bar{m}\bar{j}} = 0 \Leftrightarrow 0 = 0.$$

Thus the equation $(\Phi_\varphi\Gamma)^\gamma_{\tau\alpha\beta} = 0$ reduces to

$$\tilde{\Gamma}^k_{ij} = \Gamma^{\bar{k}}_{\bar{i}j} = \tilde{\Gamma}^{\bar{k}}_{i\bar{j}} = \Gamma^k_{ij}(x^1, \ldots, x^n), \quad \tilde{\Gamma}^{\bar{k}}_{ij} = x^{\bar{t}}\partial_t\Gamma^k_{ij} + H^k_{ij}(x^1, \ldots, x^n). \tag{3.43}$$

Remark 3.9 Using (3.31), (3.42), (3.43) and $\tilde{\Gamma}^{\gamma'}_{\alpha'\beta'} = \frac{\partial x^{\gamma'}}{\partial x^\gamma}\frac{\partial x^\alpha}{\partial x^{\alpha'}}\frac{\partial x^\beta}{\partial x^{\beta'}}\tilde{\Gamma}^\gamma_{\alpha\beta} + \frac{\partial x^{\gamma'}}{\partial x^\gamma}\frac{\partial^2 x^\gamma}{\partial x^{\alpha'}\partial x^{\beta'}}$, after straightforward calculations we see that $\Gamma^k_{ij}(x^1, \ldots, x^n)$ and $H^k_{ij}(x^1, \ldots, x^n)$ are the components of any connection ∇ and tensor field H of type (1,2) on M_n, respectively.

Taking account of the definition of the complete lift ${}^C\nabla$ of connection ∇ (see [85]), we see that a real dual-holomorphic connection $\tilde{\nabla}$ on the tangent bundle can be rewritten in the form

$$\tilde{\nabla} = {}^C\nabla + {}^VH,$$

where ${}^VH$ is the vertical lift of the tensor field $H = (H^k_{ij})$ of type (1,2) from M_n to the tangent bundle $T(M_n)$. Thus we have

Theorem 3.10 *Let $T(M_n)$ be the tangent bundle of M_n, which is the real image of the dual-holomorphic manifold $X_n(\mathbb{R}(\varepsilon))$. Then the real image of the corresponding dual-holomorphic connection from $X_n(\mathbb{R}(\varepsilon))$ is a deformed complete lift in the form ${}^D\nabla = {}^C\nabla + {}^VH$, where ${}^C\nabla$ and ${}^VH$ are the complete and vertical lifts of $\nabla = (\Gamma^k_{ij})$ and $H = (H^k_{ij})$ from M_n to $T(M_n)$, respectively.*

Example 3.1 Let (M_n, g) be a Riemannian manifold, and $(T(M_n), \varphi)$ its tangent bundle with natural dual φ-structure:

$$\varphi = \begin{pmatrix} 0 & 0 \\ I & 0 \end{pmatrix}.$$

The complete and vertical lifts of vector and tensor fields from M_n to $T(M_n)$ have the following properties

$$\varphi\,{}^CX = {}^VX, \quad {}^VX\,{}^Vf = 0, \quad {}^VX\,{}^Cf = {}^CX\,{}^Vf = {}^V(Xf),$$
$${}^Vh({}^VX, {}^CY) = 0, {}^Cg({}^VX, {}^CY) = {}^V(g(X, Y)),$$
$${}^Vh({}^CX, {}^CY) = {}^V(h(X, Y)), {}^Cg({}^CX, {}^CY) = {}^C(g(X, Y)), {}^Vh({}^VX, {}^CY) = 0,$$
$$[{}^CY, {}^VX] = {}^V[Y, X], \quad [{}^CY, {}^CX] = {}^C[Y, X]$$

for any function f on M_n [87]. Using these formulas, we find

$$\begin{aligned}
{}^{D}g(\varphi^{C}X, {}^{C}Y) &= ({}^{C}g + {}^{V}h)(\varphi\,{}^{C}X, {}^{C}Y) = ({}^{C}g + {}^{V}h)({}^{V}X, {}^{C}Y) \\
&= {}^{C}g({}^{V}X, {}^{C}Y) + {}^{V}h({}^{V}X, {}^{C}Y) = {}^{C}g({}^{V}X, {}^{C}Y) = {}^{V}(g(X, Y)) \\
&= {}^{C}g({}^{C}X, {}^{V}Y) = {}^{C}g({}^{C}X, {}^{V}Y) + {}^{V}h({}^{C}X, {}^{V}Y) = ({}^{C}g + {}^{V}h)({}^{C}X, {}^{V}Y) \\
&= ({}^{C}g + {}^{V}h)({}^{C}X, \varphi\,{}^{C}Y) \overset{Def}{=} {}^{D}g({}^{C}X, \varphi\,{}^{C}Y)
\end{aligned}$$

and

$$\begin{aligned}
(\Phi_\varphi\, {}^{D}g)({}^{C}X, {}^{C}Y, {}^{C}Z) &= (\varphi\,{}^{C}X)({}^{D}g({}^{C}Y, {}^{C}Z)) - {}^{C}X({}^{D}g(\varphi\,{}^{C}Y, {}^{C}Z)) \\
&\quad + {}^{D}g((L_{{}^{C}Y}\varphi)\,{}^{C}X, {}^{C}Z) + {}^{D}g({}^{C}Y, (L_{{}^{C}Z}\varphi)\,{}^{C}X) \\
&= {}^{V}X^{C}(g(Y, Z)) + {}^{V}X^{V}(h(Y, Z)) - {}^{C}X^{V}(g(Y, Z)) \\
&\quad + {}^{D}g(L_{{}^{C}Y}(\varphi\,{}^{C}X) - \varphi(L_{{}^{C}Y}\,{}^{C}X), {}^{C}Z) + {}^{D}g({}^{C}Y, L_{{}^{C}Z}(\varphi\,{}^{C}X) \\
&\quad - \varphi(L_{{}^{C}Z}\,{}^{C}X)) = {}^{V}(X(g(Y, Z))) - {}^{V}(X(g(Y, Z))) \\
&\quad + {}^{D}g(L_{{}^{C}Y}\,{}^{V}X - \varphi[{}^{C}Y, {}^{C}X], {}^{C}Z) + {}^{D}g({}^{C}Y, L_{{}^{C}Z}\,{}^{V}X \\
&\quad - \varphi[{}^{C}Z, {}^{C}X]) = {}^{D}g([{}^{C}Y, {}^{V}X] - \varphi^{C}[Y, X], {}^{C}Z) \\
&\quad + {}^{D}g({}^{C}Y, [{}^{C}Z, {}^{V}X] - \varphi[{}^{C}Z, {}^{C}X]) = {}^{D}g({}^{V}[Y, X] \\
&\quad - {}^{V}[Y, X], {}^{C}Z) + {}^{D}g({}^{C}Y, {}^{V}[Z, X] - {}^{V}[Z, X]) = 0.
\end{aligned}$$

From here we see that the triple $(T(M_n), {}^{D}g, \varphi)$ is a dual anti-Kähler manifold $(\nabla^{{}^{D}g}\varphi = 0)$(see Sect. 2.1). In such manifolds the Levi–Civita connection $\nabla^{{}^{D}g}$ of ${}^{D}g$ is dual-holomorphic too [25, 52]. Thus the Levi–Civita connection $\nabla^{{}^{D}g}$ is the simplest example of deformed complete lift of connection.

3.10 Holomorphic Metrics in the Tangent Bundle of Order 2

Let $\mathfrak{A}_3 = R(\varepsilon^2)$ be an algebra of order 3 with a canonical basis $\{e_1, e_2, e_3\} = \{1, \varepsilon, \varepsilon^2\}$, $\varepsilon^3 = 0$. From

$$\begin{aligned}
&C_{11}^1 = 1, C_{11}^2 = 0, C_{11}^3 = 0, C_{12}^1 = 0, C_{12}^2 = 1, C_{12}^3 = 0, \\
&C_{13}^1 = 0, C_{13}^2 = 0, C_{13}^3 = 1, C_{22}^1 = 0, C_{22}^2 = 0, C_{22}^3 = 1, \\
&C_{23}^1 = 0, C_{23}^2 = 0, C_{23}^3 = 0, C_{33}^1 = 0, C_{33}^2 = 0, C_{33}^3 = 0,
\end{aligned}$$

we see that the matrices $C_\sigma = (C_{\sigma\beta}^\gamma)$, $\sigma = 1, 2, 3$ of the regular representation of the algebra $R(\varepsilon^2)$ have the following forms

$$C_1 = \begin{pmatrix} 1\,0\,0 \\ 0\,1\,0 \\ 0\,0\,1 \end{pmatrix}, C_2 = \begin{pmatrix} 0\,0\,0 \\ 1\,0\,0 \\ 0\,1\,0 \end{pmatrix}, C_3 = \begin{pmatrix} 0\,0\,0 \\ 0\,0\,0 \\ 1\,0\,0 \end{pmatrix}$$

For the case of $\mathfrak{A}_3 = R(\varepsilon^2)$, the Scheffers conditions reduce to the following equations:

$$(i) \quad \frac{\partial y^1}{\partial x^2} = \frac{\partial y^1}{\partial x^3} = \frac{\partial y^2}{\partial x^3} = 0,$$
$$(ii) \quad \frac{\partial y^2}{\partial x^2} = \frac{\partial y^1}{\partial x^1} = \frac{\partial y^3}{\partial x^3},$$
$$(iii) \quad \frac{\partial y^3}{\partial x^2} = \frac{\partial y^2}{\partial x^1},$$

where $z = x^1 + \varepsilon x^2 + \varepsilon^2 x^3$, $w(z) = y^1(x^1, x^2, x^3) + \varepsilon y^2(x^1, x^2, x^3) + \varepsilon^2 y^3(x^1, x^2, x^3)$, $\varepsilon^3 = 0$. From (*i*) and (*ii*) we have

$$y^1 = y^1(x^1),\ y^2 = y^2(x^1, x^2),\ y^2(x^1, x^2) = x^2\frac{\mathrm{d}y^1}{\mathrm{d}x^1} + g(x^1),$$

$$y^3(x^1, x^2, x^3) = x^3\frac{\mathrm{d}y^1}{\mathrm{d}x^1} + \overline{g}(x^1, x^2)$$

After substituting the values of y^2 and y^3 into (*iii*), we find

$$\frac{\partial \overline{g}}{\partial x^2} = x^2\frac{\mathrm{d}^2 y^1}{(\mathrm{d}x^1)^2} + \frac{\mathrm{d}g}{\mathrm{d}x^1},$$

i.e.

$$\overline{g}(x^1, x^2) = \frac{1}{2}(x^2)^2\frac{\mathrm{d}^2 y^1}{(\mathrm{d}x^1)^2} + x^2\frac{\mathrm{d}g}{\mathrm{d}x^1} + G(x^1),$$

where $g = g(x^1)$ and $G = G(x^1)$ are arbitrary functions. Thus the $R(\varepsilon^2)$-holomorphic function $w = w(z)$ has the following expression

$$w(z) = y^1(x^1) + \varepsilon(x^2\frac{\mathrm{d}y^1}{\mathrm{d}x^1} + g(x^1)) + \varepsilon^2(x^3\frac{\mathrm{d}y^1}{\mathrm{d}x^1}$$
$$+ \frac{1}{2}(x^2)^2\frac{\mathrm{d}^2 y^1}{(\mathrm{d}x^1)^2} + x^2\frac{\mathrm{d}g}{\mathrm{d}x^1} + G(x^1)).$$

Similarly, if

$$w(z^1, \dots, z^n) = y^1(x^1, \dots, x^n) + \varepsilon y^2(x^1, \dots, x^2) + \varepsilon^2 y^3(x^1, \dots, x^2),$$

where $z^i = x^i + \varepsilon x^{n+i} + \varepsilon^2 x^{2n+i}$, $i = 1, \dots, n$ is a multi-variable $R(\varepsilon^2)$-holomorphic function, then the function $w = w(z^1, \dots, z^n)$ has the following specific form:

$$w(z^1, \dots, z^n) = y^1(x^1, \dots, x^n) + \varepsilon(x^{n+s}\partial_s y^1 + g(x^1, \dots, x^n))$$

$$+\,\varepsilon^2\left(x^{2n+s}\frac{\partial y^1}{\partial x^s}+\frac{1}{2}x^{n+s}x^{n+t}\frac{\partial^2 y^1}{\partial x^s\partial x^t}+x^{n+s}\frac{\partial g}{\partial x^s}+G(x^1,\ldots,x^n)\right). \tag{3.44}$$

If $g(x^1,\ldots,x^n)=G(x^1,\ldots,x^n)=0$ and $y^1(x^1,\ldots,x^n)=f(x^1,\ldots,x^n)$, then the function

$$\begin{aligned} w(z^1,\ldots,z^n) &= f(x^1,\ldots,x^n)+\varepsilon x^{n+s}\partial_s f \\ &+\varepsilon^2\left(x^{2n+s}\frac{\partial f}{\partial x^s}+\frac{1}{2}x^{n+s}x^{n+t}\frac{\partial^2 f}{\partial x^s\partial x^t}\right) \end{aligned} \tag{3.45}$$

is said to be the natural extension of the real C^∞-functions $f=f(x^1,\ldots,x^n)$ to $R(\varepsilon^2)$.

Let now $T^2(V_r)$ be the bundle of 2-jets, i.e. the tangent bundle of order 2 over the Riemannian manifold (V_r,g), $\dim T^2(V_r)=3r$ and let

$$\begin{aligned} &(x^i,x^{\bar{i}},x^{\bar{\bar{i}}})=(x^i,x^{r+i},x^{2r+i}),\, x^i=x^i(t), \\ &x^{\bar{i}}=\frac{\mathrm{d}x^i}{\mathrm{d}t},\, x^{\bar{\bar{i}}}=\frac{1}{2}\frac{\mathrm{d}^2x^i}{\mathrm{d}t^2},\quad t\in\mathbb{R},\quad i=1,\ldots,r \end{aligned}$$

be an induced local coordinates system in $T^2(V_r)$. It is clear that there exists an affinor field γ in $T^2(V_r)$ which has the components of the form

$$\gamma=\begin{pmatrix}0&0&0\\ I&0&0\\ 0&I&0\end{pmatrix}$$

with respect to the natural frame $\{\partial_i,\partial_{\bar{i}},\partial_{\bar{\bar{i}}}\}=\left\{\frac{\partial}{\partial x^i},\frac{\partial}{\partial x^{\bar{i}}},\frac{\partial}{\partial x^{\bar{\bar{i}}}}\right\}$, $i=1,\ldots,r$, where I denotes the $r\times r$ identity matrix. From here we have

$$\gamma^2=\begin{pmatrix}0&0&0\\ 0&0&0\\ I&0&0\end{pmatrix},\quad \gamma^3=0,$$

i.e. $T^2(V_r)$ has a natural integrable structure $\Pi=\{I,\gamma,\gamma^2\}$, $I=id_{T^2(V_r)}$, which is an isomorphic representation of the algebra $R(\varepsilon^2)$, $\varepsilon^3=0$. Using

$$\gamma\partial_i=\partial_{\bar{i}},\quad \gamma^2\partial_i=\gamma\partial_{\bar{i}}=\partial_{\bar{\bar{i}}},$$

we have $\{\partial_i,\partial_{\bar{i}},\partial_{\bar{\bar{i}}}\}=\{\partial_i,\gamma\partial_i,\gamma^2\partial_i\}$. Also, using a frame

$$\{\partial_1,\gamma\partial_1,\gamma^2\partial_1,\partial_2,\gamma\partial_2,\gamma^2\partial_2,\ldots,\partial_r,\gamma\partial_r,\gamma^2\partial_r\}=\{\partial_1,\partial_{\bar{1}},\partial_{\bar{\bar{1}}},\partial_2,\partial_{\bar{2}},\partial_{\bar{\bar{2}}},\ldots,\partial_r,\partial_{\bar{r}},\partial_{\bar{\bar{r}}}\}$$

which is obtained from $\{\partial_i, \partial_{\overline{i}}, \partial_{\overline{\overline{i}}}\} = \{\partial_i, \gamma\partial_i, \gamma^2\partial_i\}$ by changing the numbers of frame elements, we see that the structure affinors I, γ and γ^2 have the following components

$$I = \begin{pmatrix} C_1 & 0 & \dots & 0 \\ 0 & C_1 & \dots & 0 \\ \dots & \dots & \dots & \dots \\ 0 & 0 & \dots & C_1 \end{pmatrix}, \quad \gamma = \begin{pmatrix} C_2 & 0 & \dots & 0 \\ 0 & C_2 & \dots & 0 \\ \dots & \dots & \dots & \dots \\ 0 & 0 & \dots & C_2 \end{pmatrix}, \quad \gamma^2 = \begin{pmatrix} C_3 & 0 & \dots & 0 \\ 0 & C_3 & \dots & 0 \\ \dots & \dots & \dots & \dots \\ 0 & 0 & \dots & C_3 \end{pmatrix}$$

with respect to the frame $\{\partial_1, \partial_{\overline{1}}, \partial_{\overline{\overline{1}}}, \partial_2, \partial_{\overline{2}}, \partial_{\overline{\overline{2}}}, \dots, \partial_r, \partial_{\overline{r}}, \partial_{\overline{\overline{r}}}\}$, where the block matrices $C_\sigma, \sigma = 1, 2, 3$ of order 3 are the regular representation of the algebra $R(\varepsilon^2)$. Thus the bundle $T^2(V_r)$ has a natural integrable structure $\Pi = \{I, \gamma, \gamma^2\}$, which is an r-regular representation of $R(\varepsilon^2)$.

On the other hand, the transformation of induced coordinates $(x^i, x^{\overline{i}}, x^{\overline{\overline{i}}})$ in $T^2(V_r)$ is given by

$$\begin{aligned} x^{i'} &= x^{i'}(x^i), \\ x^{\overline{i}'} &= \frac{dx^{i'}}{dt} = \frac{\partial x^{i'}}{\partial x^i}\frac{dx^i}{dt} = \frac{\partial x^{i'}}{\partial x^i}x^{\overline{i}}, \\ x^{\overline{\overline{i}}'} &= \frac{1}{2}\frac{d^2x^{i'}}{dt^2} = \frac{1}{2}\frac{d}{dt}\left(\frac{\partial x^{i'}}{\partial x^i}\frac{dx^i}{dt}\right) = \frac{1}{2}\frac{\partial x^{i'}}{\partial x^i}\frac{d^2x^i}{dt^2} + \frac{1}{2}\frac{\partial^2 x^{i'}}{\partial x^i \partial x^j}\frac{dx^i}{dt}\frac{dx^j}{dt} \\ &= \frac{\partial x^{i'}}{\partial x^i}x^{\overline{\overline{i}}} + \frac{1}{2}\frac{\partial^2 x^{i'}}{\partial x^i \partial x^j}x^{\overline{i}}x^{\overline{j}} \end{aligned}$$

and its Jacobian matrix by

$$A = \begin{pmatrix} \frac{\partial x^{i'}}{\partial x^i} & \frac{\partial x^{i'}}{\partial x^{\overline{i}}} & \frac{\partial x^{i'}}{\partial x^{\overline{\overline{i}}}} \\ \frac{\partial x^{\overline{i}'}}{\partial x^i} & \frac{\partial x^{\overline{i}'}}{\partial x^{\overline{i}}} & \frac{\partial x^{\overline{i}'}}{\partial x^{\overline{\overline{i}}}} \\ \frac{\partial x^{\overline{\overline{i}}'}}{\partial x^i} & \frac{\partial x^{\overline{\overline{i}}'}}{\partial x^{\overline{i}}} & \frac{\partial x^{\overline{\overline{i}}'}}{\partial x^{\overline{\overline{i}}}} \end{pmatrix} = \begin{pmatrix} \frac{\partial x^{i'}}{\partial x^i} & 0 & 0 \\ \frac{\partial^2 x^{i'}}{\partial x^i \partial x^s}x^{\overline{s}} & \frac{\partial x^{i'}}{\partial x^i} & 0 \\ \frac{\partial^2 x^{i'}}{\partial x^i \partial x^s}x^{\overline{\overline{s}}} + \frac{\partial^3 x^{i'}}{\partial x^i \partial x^s \partial x^t}x^{\overline{s}}x^{\overline{t}} & \frac{\partial^2 x^{i'}}{\partial x^i \partial x^s}x^{\overline{s}} & \frac{\partial x^{i'}}{\partial x^i} \end{pmatrix}. \tag{3.46}$$

From here it follows that $A^{-1}\gamma A = \gamma, \; A^{-1}\gamma^2 A = \gamma^2$, i.e. the transformation of local coordinates $(x^i, x^{\overline{i}}, x^{\overline{\overline{i}}})$ in $T^2(V_r)$ is a structure-preserving transformation. Then the transition functions

$$z^{i'}(z^i) = x^{i'} + \varepsilon x^{\overline{i}'} + \varepsilon^2 x^{\overline{\overline{i}}'} = x^{i'}(x^i) + \varepsilon\frac{\partial x^{i'}}{\partial x^i}x^{\overline{i}} + \varepsilon^2\left(\frac{\partial x^{i'}}{\partial x^i}x^{\overline{\overline{i}}} + \frac{1}{2}\frac{\partial^2 x^{i'}}{\partial x^i \partial x^j}x^{\overline{i}}x^{\overline{j}}\right)$$

of charts on $X_r(R(\varepsilon^2))$ are $R(\varepsilon^2)$-holomorphic functions by virtue of (3.45), i.e. the bundle $T^2(V_r)$ is a real modeling of the $R(\varepsilon^2)$-holomorphic manifold $X_r(R(\varepsilon^2))$.

After some calculations, we see that the 2-nd lift ${}^{II}g$ [87] (see p. 332) of a Riemannian metric g to $T^2(V_r)$, i.e.

$$^{II}g = \begin{pmatrix} x^{\bar{s}}\partial_s g_{ji} + \frac{1}{2}x^{\bar{t}}x^{\bar{s}}\partial_t\partial_s g_{ji} & x^{\bar{s}}\partial_s g_{ji} & g_{ji} \\ x^{\bar{s}}\partial_s g_{ji} & g_{ji} & 0 \\ g_{ji} & 0 & 0 \end{pmatrix}$$

is a pure Riemannian metric with respect to the structure $\Pi = \{I, \gamma, \gamma^2\}$ and

$$\Phi_\gamma {}^{II}g = \Phi_{\gamma^2} {}^{II}g = 0.$$

Therefore, the $R(\varepsilon^2)$-holomorphic manifold $(T^2(V_r), \Pi)$ is a manifold with $R(\varepsilon^2)$-holomorphic metric ^{II}g. From here, using Theorem 2.1, we have

Theorem 3.11 [76] *If V_r is a Riemannian manifold with metric g, then the triple $(T^2(V_r), \Pi, {}^{II}g)$ is an anti-Kähler type manifold.*

Finally, we would like to find the local expression of any $R(\varepsilon^2)$-holomorphic pure tensor field $\tilde{t}$ of type (0,2) in $T^2(V_r)$. Using

$$\tilde{t}_{MJ}\gamma_I^M = \tilde{t}_{IM}\gamma_J^M, \quad I = (i, \bar{i}, \bar{\bar{i}}), \quad M = (m, \bar{m}, \bar{\bar{m}}), \quad J = (j, \bar{j}, \bar{\bar{j}})$$

and

$$(\Phi_\gamma \tilde{t})_{KIJ} = \gamma_K^M \partial_M \tilde{t}_{IJ} - \gamma_I^M \partial_K \tilde{t}_{MJ} = 0,$$
$$(\Phi_{\gamma^2} \tilde{t})_{KIJ} = (\gamma^2)_K^M \partial_M \tilde{t}_{IJ} - (\gamma^2)_I^M \partial_K \tilde{t}_{MJ} = 0,$$

after straightforward calculations, we find

$$\tilde{t} = \begin{pmatrix} x^{\bar{s}}\partial_s t_{ji} + \frac{1}{2}x^{\bar{t}}x^{\bar{s}}\partial_t\partial_s t_{ji} + x^{\bar{s}}\partial_s G_{ji} + H_{ji} & x^{\bar{s}}\partial_s t_{ji} + G_{ji} & t_{ji} \\ x^{\bar{s}}\partial_s t_{ji} + G_{ji} & t_{ji} & 0 \\ t_{ji} & 0 & 0 \end{pmatrix},$$

where t_{ji}, G_{ji}, H_{ji} are arbitrary tensor fields of type (0,2) in V_r. If $t_{ij} = g_{ij}$ and G_{ji}, H_{ji} are symmetric tensor fields in V_r, then we have a new $R(\varepsilon^2)$-holomorphic pure Riemannian metric in $T^2(V_r)$:

$$\tilde{g} = \begin{pmatrix} x^{\bar{s}}\partial_s g_{ji} + \frac{1}{2}x^{\bar{t}}x^{\bar{s}}\partial_t\partial_s g_{ji} + x^{\bar{s}}\partial_s G_{ji} + H_{ji} & x^{\bar{s}}\partial_s g_{ji} + G_{ji} & g_{ji} \\ x^{\bar{s}}\partial_s g_{ji} + G_{ji} & g_{ji} & 0 \\ g_{ji} & 0 & 0 \end{pmatrix}. \tag{3.47}$$

We denote $\tilde{g}$ by $^{def}(^{II}g)$ and call it a deformed 2-nd lift of a Riemannian metric g to $T^2(V_r)$. By using again Theorem 2.1, we have

Theorem 3.12 *If V_r is a Riemannian manifold with metric g, then the triple $(T^2(V_r), \Pi, {}^{def}({}^{II}g))$ is an anti-Kähler type manifold.*

From (3.47) we see that a general form of the deformed complete lift ${}^{def}({}^{II}g)$ is

$$ {}^{def}({}^{II}g) = {}^{II}g + {}^{I}G + {}^{0}H, $$

where

$$ {}^{I}G = \begin{pmatrix} x^{\bar{s}}\partial_s G_{ji} & G_{ji} & 0 \\ G_{ji} & 0 & 0 \\ 0 & 0 & 0 \end{pmatrix} \text{ and } {}^{0}H = \begin{pmatrix} H_{ji} & 0 & 0 \\ 0 & 0 & 0 \\ 0 & 0 & 0 \end{pmatrix} $$

are the 1-st and 0-th lifts [87] of G and H, respectively.

3.11 Deformed Lifts of Vector Fields in the Tangent Bundle of Order 2

Let now $T^2(M_r)$ be the tangent bundle of order 2 over a C^∞-manifold M_r, $\dim T^2(M_r) = 3r$ and lct

$$ (x^i, x^{\bar{i}}, x^{\bar{\bar{i}}}) = (x^i, x^{r+i}, x^{2r+i}), x^i = x^i(t), x^{\bar{i}} = \frac{dx^i}{dt}, x^{\bar{\bar{i}}} = \frac{1}{2}\frac{d^2x^i}{dt^2}, t \in \mathbb{R}, i = 1, \dots, r $$

be the induced local coordinates in $T^2(M_r)$. Since the bundle $T^2(M_r)$ is a real modeling of $X_r(R(\varepsilon^2))$ and any holomorphic function

$$ w(z^1, \dots, z^r) = f^1(x^1, \dots, x^r) + \varepsilon f^2(x^1, \dots, x^r) + \varepsilon^2 f^3(x^1, \dots, x^r) $$

on $X_r(R(\varepsilon^2))$, where $z^i = x^i + \varepsilon x^{r+i} + \varepsilon^2 x^{2r+i}$, $i = 1, \dots, r$, is expressed by (see (3.44))

$$ \begin{aligned} w(z^1, \dots, z^r) = {} & f(x^1, \dots, x^r) + \varepsilon(x^{r+i}\partial_i f + g(x^1, \dots, x^r)) \\ & + \varepsilon^2(x^{2r+i}\frac{\partial f}{\partial x^i} + \frac{1}{2}x^{r+i}x^{r+j}\frac{\partial^2 f}{\partial x^i \partial x^j} \\ & + x^{r+i}\frac{\partial g}{\partial x^i} + h(x^1, \dots, x^r)), \quad f = f^1, \end{aligned} $$

in the bundle $T^2(M_r)$, we introduce the following three functions:

$$ \begin{aligned} & {}^{V}f = f(x^1, \dots, x^r), \\ & {}^{I}f = x^{r+i}\partial_i f + g(x^1, \dots, x^r), \end{aligned} $$

$$
{}^{C}f = x^{2r+i}\frac{\partial f}{\partial x^i} + \frac{1}{2}x^{r+i}x^{r+j}\frac{\partial^2 f}{\partial x^i \partial x^j} + x^{r+i}\frac{\partial g}{\partial x^i} + h(x^1, \ldots, x^r),
$$

where f, g and h are arbitrary functions on M_r. These functions ${}^{V}f, {}^{I}f$ and ${}^{C}f$ are called recpectively the *vertical, intermediate and complete* lifts of f in M_r to $T^2(M_r)$. If $g = h = 0$, then we have the *0-th* f^0, *1-th* f^1 *and 2-nd* f^2 lifts of f [87], i.e. the lifts ${}^{I}f$ and ${}^{C}f$ of f to $T^2(M_r)$ are respectively the *deformed lifts* of 1-th and 2-nd lifts of f. Thus we have

$$
{}^{V}f = f^0, \quad {}^{I}f = f^1 + g^0, \quad {}^{C}f = f^2 + g^1 + h^0. \tag{3.48}
$$

We now consider in $T^2(M_r)$ two vector fields of *Liouville types* U and V having components

$$
U = \begin{pmatrix} 0 \\ 0 \\ x^{r+i} \end{pmatrix}, \quad V = \begin{pmatrix} 0 \\ x^{r+i} \\ x^{2r+i} \end{pmatrix}
$$

Then from (3.48), we immediately have

Theorem 3.13 *Let U and V be the Liouville vector fields in $T^2(M_r)$. Then*

$$
U({}^{V}f) = 0, U({}^{I}f) = 0, U({}^{C}f) = f^1, V({}^{V}f) = 0, V({}^{I}f) = f^1, V({}^{C}f) = f^2 + g^1
$$

for any function f in M_r.

Let $\tilde{X} = \tilde{X}^I\frac{\partial}{\partial x^I} = \tilde{X}^i\frac{\partial}{\partial x^i} + \tilde{X}^{r+i}\frac{\partial}{\partial x^{r+i}} + \tilde{X}^{2r+i}\frac{\partial}{\partial x^{2r+i}}$ be a vector field in $T^2(M_r)$, and $\Pi = \{I, \gamma, \gamma^2\}, I = id_{T^2(M_r)}$ be a Π-structure naturally existing in $T^2(M_r)$. We would like to find the local expression of any vector field $\tilde{X} = (\tilde{X}^I)$ in $T^2(M_r)$ corresponding to the $R(\varepsilon^2)$-holomorphic vector field $\overset{*}{X} = (\overset{*}{X}{}^u) = (X^{u\alpha}e_\alpha), \quad u = 1, \ldots, r; \alpha = 1, 2, 3$ in $X_r(R(\varepsilon^2))$. Using Theorem 1.9, we obtain

$$
\begin{aligned}
(\Phi_\gamma \tilde{X})^I &= \gamma_K^M \partial_M \tilde{X}^I - \gamma_M^I \partial_K \tilde{X}^M = 0, \\
(\Phi_{\gamma^2} \tilde{X})^I &= (\gamma^2)_K^M \partial_M \tilde{X}^I - (\gamma^2)_M^I \partial_K \tilde{X}^M = 0.
\end{aligned}
$$

From here, after straightforward calculations (see Sect. 3.10), we find the following vector field

$$
{}^{C}X = \begin{pmatrix} X^h \\ x^{r+i}\partial_i X^h + G^h \\ x^{2r+i}\frac{\partial X^h}{\partial x^i} + \frac{1}{2}x^{r+i}x^{r+j}\frac{\partial^2 X^h}{\partial x^i \partial x^j} + x^{r+i}\frac{\partial G^h}{\partial x^i} + H^h \end{pmatrix}, \tag{3.49}
$$

where $G = (G^h(x^1, \dots, x^r))$, $H = (H^h(x^1, \dots, x^r))$ are two arbitrary vector fields in M_r. In fact, by means of (3.46), we easily see that ${}^C X$ determine some vector fields in $T^2(M_r)$, which are called the *complete lifts* of X from M_r to $T^2(M_r)$. From (3.49), we have

$$ {}^C X = X^2 + G^1 + H^0, \tag{3.50}$$

where

$$H^0 = \begin{pmatrix} 0 \\ 0 \\ H^h \end{pmatrix}, G^1 = \begin{pmatrix} 0 \\ G^h \\ x^{r+i}\partial_i G^h \end{pmatrix}, X = \begin{pmatrix} X^h \\ x^{r+i}\partial_i X^h \\ x^{2r+i}\frac{\partial X^h}{\partial x^i} + \frac{1}{2}x^{r+i}x^{r+j}\frac{\partial^2 X^h}{\partial x^i \partial x^j} \end{pmatrix}$$

are respectively the *0-th, 1-th and 2-nd* lifts of H, G and X [87]. Putting $G = X$, $H = G$ and using (3.50), we see that

$$ {}^I X = X^1 + G^0 = {}^C X - X^2 = \begin{pmatrix} 0 \\ X^h \\ x^{r+i}\partial_i X^h + G^h \end{pmatrix}, \tag{3.51}$$

determines a new vector field in $T^2(M_r)$, which is called the *intermediate lift* of a vector field X from M_r to $T^2(M_r)$. We note that the intermediate and complete lifts ${}^I X$ and ${}^C X$ of X to $T^2(M_r)$ are respectively the *deformed lifts* of 1-th and 2-th lifts of X. Thus we have

$$ {}^V X = X^0, \quad {}^I X = X^1 + G^0, \quad {}^C X = X^2 + G^1 + H^0, \tag{3.52}$$

where

$$ {}^V X = \begin{pmatrix} 0 \\ 0 \\ X^h \end{pmatrix}. \tag{3.53}$$

Using the local expressions of γ and γ^2 (see Sect. 3.10) and also (3.49), (3.51) and (3.53) we have

$$ {}^I X = \gamma({}^C X), \quad {}^V X = \gamma^2({}^C X).$$

In particular, for the vector fields $X = \partial_i$, $G = \partial_j$, $H = \partial_k$, from (3.49) to (3.53) we obtain

$$ {}^V(\partial_i) = \partial_{2r+i}, {}^I(\partial_i) = \partial_{r+i} + \partial_{2r+j}, {}^C(\partial_i) = \partial_i + \partial_{r+j} + \partial_{2r+k}.$$

By (3.48), (3.49) and (3.50) we have

Theorem 3.14 *Let X, G, H and f, g be respectively any vector fields and functions in M_r. Then*

$$
\begin{aligned}
&{}^V(fX) = f^0X^0, {}^I(fX) = f^1X^0 + f^0X^1 + G^0,\\
&{}^C(fX) = f^2X^0 + f^1X^1 + f^0X^0 + G^1 + H^0,\\
&{}^VX\,{}^Vf = 0, {}^VX\,{}^If = 0, {}^VX\,{}^Cf = (Xf)^0,\\
&{}^IX\,{}^Vf = 0, {}^IX\,{}^If = (Xf)^0, {}^IX\,{}^Cf = (Xf)^1 + (Xg)^0 + (Gf)^0,\\
&{}^CX\,{}^Vf = (Xf)^0, {}^CX^If = (Xf)^1 + (Xg)^0 + (Gf)^0,\\
&{}^CX\,{}^Cf = (Xf)^2 + (Xg)^1 + (Xf)^0 + (Gf)^1 + (Gg)^0 + (Hf)^0.
\end{aligned}
$$

Theorem 3.15 *Let X, Y, G, Q, H, K be any vector vector fields in M_r. Then for the Lie bracket of complete lifts we have*

$$[{}^CX, {}^CY] = [X, Y]^2 + [X, Q]^1 + [X, K]^0 + [G, Y]^1 + [G, Q]^0 + [H, Y]^0,$$

where

$${}^CX = X^2 + G^1 + H^0, \quad {}^CY = Y^2 + Q^1 + K^0.$$

Proof If f is any function in M_r, then by virtue of (3.52), we have

$$
\begin{aligned}
[{}^CX, {}^CY](f^2) = {} & [X^2, Y^2](f^2) + [X^2, Q^1](f^2) + [X^2, K^0](f^2)\\
& + [G^1, Y^2](f^2) + [G^1, Q^1](f^2) + [G^1, K^0](f^2)\\
& + [H^0, Y^2](f^2) + [H^0, Q^1](f^2) + [H^0, K^0](f^2)
\end{aligned}
$$

Using the formulas (see [87, p. 322])

$$
\begin{aligned}
&[X^2, Y^2] = [X, Y]^2, [X^2, Q^1] = [X, Q]^1, [X^2, K^0] = [X, K]^0,\\
&[G^1, Y^2] = [G, Y]^1, [G^1, Q^1] = [G, Q]^0, [G^1, K^0] = 0, [H^0, Y^2] = [H, Y]^0,\\
&[H^0, Q^1] = 0, [H^0, K^0] = 0
\end{aligned}
$$

we have

$$
\begin{aligned}
[{}^CX, {}^CY](f^2) = {} & [X, Y]^2(f^2) + [X, Q]^1(f^2) + [X, K]^0(f^2) + [G, Y]^1(f^2)\\
& + [G, Q]^0(f^2) + [H, Y]^0(f^2) = ([X, Y]^2 + [X, Q]^1 + [X, K]^0 + [G, Y]^1\\
& + [G, Q]^0 + [H, Y]^0)(f^2).
\end{aligned}
$$

Since the vector field $\tilde{X}$ in $T^2(M_r)$ is completely determined by the action of $\tilde{X}$ on the functions f^2 (see [87, p. 320]), i.e. if $\tilde{X}f^2 = \tilde{Y}f^2$ for any f, then $\tilde{X} = \tilde{Y}$, we have

$$[{}^C X, {}^C Y] = [X, Y]^2 + [X, Q]^1 + [X, K]^0 + [G, Y]^1 + [G, Q]^0 + [H, Y]^0.$$

Thus the proof is completed.

From Theorem 3.15 we have

$$\begin{aligned}[{}^C X, {}^C Y] &= [X, Y]^2 + [X, Q]^1 + [X, K]^0 + [G, Y]^1 + [G, Q]^0 + [H, Y]^0 \\ &= {}^C[X, Y] + [X, Q]^1 + [X, K]^0 + [G, Y]^1 + [G, Q]^0 + [H, Y]^0 \\ &\quad - [X, Y]^1 - [X, Y]^0,\end{aligned}$$

i.e. the correspondence $X \to {}^C X$ is not an isomorphism of the Lie algebras of vector fields on M_r and $T^2(M_r)$ with respect to constant coefficients.

3.12 Deformed Complete and Intermediate Lifts of 1-Forms in the Tangent Bundle of Order 2

Let $\tilde{\omega} = \tilde{\omega}_I dx^I = \tilde{\omega}_i dx^i + \tilde{\omega}_{r+i} dx^{r+i} + \tilde{\omega}_{2r+i} dx^{2r+i}$ be an 1-form in $T^2(M_r)$, and $\Pi = \{I, \gamma, \gamma^2\}$, $I = id_{T^2(M_r)}$ be a Π-structure naturally existing in $T^2(M_r)$. We would like to find local expression of $\tilde{\omega} = (\tilde{\omega}_I)$ in $T^2(M_r)$ which is corresponding to the $R(\varepsilon^2)$-holomorphic 1-form $\overset{*}{\omega} = (\overset{*}{\omega}_u) = (\overset{*}{\omega}_{u\alpha} c^\alpha)$, $c^\alpha = \varphi^{\alpha\beta} c_\beta$, $u = 1, \ldots, r$; $\alpha, \beta = 1, 2, 3$ in $X_r(R(\varepsilon^2))$. Using Theorem 1.9, we obtain

$$\begin{aligned}(\Phi_\gamma \tilde{\omega})_{JI} &= \gamma_J^H \partial_H \tilde{\omega}_I - \gamma_I^H \partial_J \tilde{\omega}_H = 0, \\ (\Phi_{\gamma^2} \tilde{\omega})_{JI} &= (\gamma^2)_J^H \partial_H \tilde{\omega}_I - (\gamma^2)_I^H \partial_J \tilde{\omega}_H = 0.\end{aligned}$$

From here, after straightforward calculations, we find

$$\tilde{\omega} = (\tilde{\omega}_I) = (x^{2r+h} \partial_h \omega_i + \frac{1}{2} x^{r+h} x^{r+m} \partial^2_{hm} \omega_i + x^{h+i} \partial_h G_i + H_i, x^{r+h} \partial_h \omega_i + G_i, \omega_i), \tag{3.54}$$

where $G = (G_i(x^1, \ldots, x^r))$, $H = (H_i(x^1, \ldots, x^r))$ are any covector fields in M_r. In fact, by means of (3.46), we easily see that $\tilde{\omega} = (\tilde{\omega}_I)$ determine the 1-form in $T^2(M_r)$ which are called the *deformed complete lifts* of ω from M_r to $T^2(M_r)$ and denoted by ${}^C\omega = ({}^C\omega_I)$. From (3.54), we have

$${}^C\omega = \omega^2 + G^1 + H^0, \tag{3.55}$$

where

$$\begin{aligned}H^0 &= (H_i, 0, 0), \\ G^1 &= (x^{r+h} \partial_h G_i, G_i, 0),\end{aligned}$$

$$\omega^2 = (x^{2r+h}\partial_h\omega_i + \frac{1}{2}x^{r+h}x^{r+m}\partial^2_{hm}\omega_i, x^{r+h}\partial_h\omega_i, \omega_i)$$

are respectively the *0-th (vertical), 1-th and 2-nd (complete) lifts* of H, G and ω [87]. Thus we have

Theorem 3.16 *Let $\omega = \omega_i \mathrm{d}x^i$ be a 1-form on M_r. The deformed complete lift ${}^C\omega$ of ω to the bundle of 2-jets $T^2(M_r)$ have the following expression*

$$^C\omega = \omega^2 + G^1 + H^0,$$

where H^0, G^1 and ω^2 are respectively the 0-th, 1-th and 2-th lifts of any 1-forms H, G and ω.

Putting $\omega = G$ in (3.55), we see that

$$\omega^1 + H^0 = {}^C\omega - \omega^2 = (x^{r+i}\partial_i\omega_h + H_h, \omega_h, 0) \tag{3.56}$$

determine a new 1-form in $T^2(M_r)$, which are called the *deformed intermediate lift* of 1-form ω from M_r to $T^2(M_r)$ and denoted by ${}^I\omega = \omega^1 + H^0$. We note that the *deformed intermediate lift* ${}^I\omega$ of ω to $T^2(M_r)$ is deformation of 1-th lift of ω. Thus we have

$$^V\omega = \omega^0, \quad {}^I\omega = \omega^1 + H^0, \quad {}^C\omega = \omega^2 + G^1 + H^0, \tag{3.57}$$

where

$$^V\omega = (\omega_h, 0, 0). \tag{3.58}$$

Now we can state the following

Theorem 3.17 *Let $\omega = \omega_i \mathrm{d}x^i$ be a 1-form on M_r. The deformed intermediate lift ${}^I\omega$ of ω to the bundle of 2-jets $T^2(M_r)$ have the following expression*

$$^I\omega = \omega^1 + H^0,$$

where H^0 is the 0-th lift of 1-form H.

Using the local expressions of γ and γ^2 (see Sect. 3.10), from (3.54), (3.56) and (3.58) we have

Theorem 3.18 *The deformed complete lifts satisfy the following matrix formulas*

$$^C\omega\gamma = \omega^1 + G^0, \quad {}^C\omega\gamma^2 = \omega^0 = {}^V\omega,$$

where

$$\gamma = \begin{pmatrix} 0\,0\,0 \\ I\,0\,0 \\ 0\,I\,0 \end{pmatrix}, \quad \gamma^2 = \begin{pmatrix} 0\,0\,0 \\ 0\,0\,0 \\ I\,0\,0 \end{pmatrix}.$$

Let now $\omega = \mathrm{d}x^i, G = \mathrm{d}x^j, H = \mathrm{d}x^k, i, j, k = 1, \ldots, r$. Then from (3.56) to (3.58) we have

Theorem 3.19 *Deformed complete, intermediate and vertical lifts of differentials* $\mathrm{d}x^i$ *are the following linear combinations of differentials in* $T^2(M_r)$:

$$^V(\mathrm{d}x^i) = \mathrm{d}x^i, {}^I(\mathrm{d}x^i) = \mathrm{d}x^{r+i} + \mathrm{d}x^k, {}^C(\mathrm{d}x^i) = \mathrm{d}x^{2r+i} + \mathrm{d}x^{r+j} + \mathrm{d}x^k.$$

Let now X be a vector field in M_r. It is well known that the vertical and deformed lifts ${}^I X, {}^C X$ of X have the following expressions (see Sect. 3.11):

$$^V X = X^0 = \begin{pmatrix} 0 \\ 0 \\ X^h \end{pmatrix},$$

$$^I X = X^1 + Y^0 = \begin{pmatrix} 0 \\ X^h \\ x^{r+i}\partial_i X^h \end{pmatrix} + \begin{pmatrix} 0 \\ 0 \\ Y^h \end{pmatrix} \begin{pmatrix} 0 \\ X^h \\ x^{r+i}\partial_i X^h + Y^h \end{pmatrix},$$

$$^C X = X^2 + Y^1 + Z^0 = \begin{pmatrix} X^h \\ x^{r+i}\partial_i X^h \\ x^{2r+i}\partial_i X^h + \frac{1}{2}x^{r+i}x^{r+j}\partial^2_{ij} X^h \end{pmatrix} + \begin{pmatrix} 0 \\ Y^h \\ x^{r+i}\partial_i Y^h \end{pmatrix} + \begin{pmatrix} 0 \\ 0 \\ Z^h \end{pmatrix}$$

$$= \begin{pmatrix} X^h \\ x^{r+i}\partial_i X^h + Y^h \\ x^{2r+i}\partial_i X^h + \frac{1}{2}x^{r+i}x^{r+j}\partial^2_{ij} X^h + x^{r+i}\partial_i Y^h + Z^h \end{pmatrix}$$

for any vector fields Y, Z in M_r. Using the last formulas and also (3.48), (3.56)–(3.58) we have

Theorem 3.20 *Let* X, ω *and* f *be respectively any vector field, 1-form and function in* M_r. *Then*

$$^V(f\omega) = f^0\omega^0, {}^I(f\omega) = f^1\omega^0 + f^0\omega^1 + G^0, {}^C(f\omega) = (f^2 + f^0)\omega^0 + f^1\omega^1 + G^1 + H^0,$$
$$^V\omega({}^V X) = 0, {}^V\omega({}^I X) = 0, {}^V\omega({}^C X) = (\omega(X))^0,$$
$$^I\omega({}^V X) = 0, {}^I\omega({}^I X) = (\omega(X))^0, {}^I\omega({}^C X) = (\omega(X))^1 + (\omega(Y))^0 + (H(X))^0,$$
$$^C\omega({}^V X) = (\omega(X))^0, {}^C\omega({}^I X) = (\omega(X))^1 + (\omega(Y))^0 + (G(X))^0,$$

$$^{C}\omega(^{C}X) = (\omega(X))^2 + (\omega(Y))^1 + (\omega(Z))^0 + (G(X))^1 + (G(Y))^0 + (H(X))^0.$$

Let now Ω be a tensor field of type (0,2) in M_r. We define a 1-form $\gamma_Y\Omega$ by

$$(\gamma_Y\Omega)X = \Omega(X, Y)$$

for any vector fields X and Y. If Ω has the local components Ω_{ij}, then $\gamma_Y\Omega$ has the local components $\Omega_{ij}Y^j$.

It is well known that the deformed intermediate and complete lifts of Ω has respectively components (see Sect. 3.10)

$$^{I}\Omega = \begin{pmatrix} x^{r+s}\partial_s\Omega_{ji} + \pi_{ji} & \Omega_{ji} & 0 \\ \Omega_{ji} & 0 & 0 \\ 0 & 0 & 0 \end{pmatrix} = \Omega^1 + \pi^0,$$

$$^{C}\Omega = \begin{pmatrix} x^{2r+s}\partial_s\Omega_{ji} + \frac{1}{2}x^{r+t}x^{r+s}\partial^2_{ts}\Omega_{ji} + x^{r+s}\partial_s\Omega_{ji} + \pi_{ji} & x^{r+s}\partial_s\Omega_{ji} + \Omega_{ji} & \Omega_{ji} \\ x^{r+s}\partial_s\Omega_{ji} + \Omega_{ji} & \Omega_{ji} & 0 \\ \Omega_{ji} & 0 & 0 \end{pmatrix}$$
$$= \Omega^2 + \Omega^1 + \pi^0,$$

where

$$^{V}\pi = {}^{0}\pi = \begin{pmatrix} \pi_{ji} & 0 & 0 \\ 0 & 0 & 0 \\ 0 & 0 & 0 \end{pmatrix}$$

is the vertical lift of any tensor field π of type (0,2). Using the expression of lifts $^{V}\pi$, $^{I}\Omega$, $^{C}\Omega$ and (3.54), (3.56)–(3.58) we have

$$\begin{aligned}
\gamma_{X^2}\, {}^{V}\Omega &= (\Omega_{ij}X^j, 0, 0) = ((\gamma_X\Omega)_i, 0, 0) = {}^{V}(\gamma_X\Omega),\\
\gamma_{X^2}\, {}^{I}\Omega &= ((x^{r+s}\partial_s\Omega_{ij} + \pi_{ij})X^j + \Omega_{ij}(x^{r+s}\partial_s X^j), \Omega_{ij}X^j, 0)\\
&= (x^{r+s}\partial_s(\gamma_X\Omega)_i + (\gamma_X\pi)_i,\\
(\gamma_X\Omega)_i, 0) &= (\gamma_X\Omega)^1 + (\gamma_X\pi)^0 = {}^{I}(\gamma_X\Omega),\\
\gamma_{X^2}\, {}^{C}\Omega &= ((x^{2r+s}\partial_s\Omega_{ij} + \frac{1}{2}x^{r+s}x^{r+t}\partial^2_{st}\Omega_{ij} + x^{r+s}\partial_s\Omega_{ij} + \pi_{ij})X^j\\
&+ (x^{r+s}\partial_s\Omega_{ij} + \Omega_{ij})x^{r+t}\partial_t X^j\\
&+ \Omega_{ij}(x^{2r+s}\partial_s X^j + \frac{1}{2}x^{r+s}x^{r+t}\partial^2_{st}X^j, (x^{r+s}\partial_s\Omega_{ij} + \Omega_{ij})X^j\\
&+ \Omega_{ij}x^{r+t}\partial_t X^j, \Omega_{ij}X^j)\\
&= (x^{2r+s}\partial_s(\gamma_X\Omega)_i + \frac{1}{2}x^{r+s}x^{r+t}\partial^2_{st}(\gamma_X\Omega)_i + x^{r+s}\partial_s(\gamma_X\Omega)_i
\end{aligned}$$

$$+ (\gamma_X\pi)_i, x^{r+s}\partial_s(\gamma_X\Omega)_i + (\gamma_X\Omega)_i, (\gamma_X\Omega)_i)$$
$$= (\gamma_X\Omega)^2 + (\gamma_X\Omega)^1 + (\gamma_X\pi)^0 = {}^C(\gamma_X\Omega).$$

Thus we have

Theorem 3.21 *Let Ω be a tensor field of type (0,2) on M_r. Then*

$$\gamma_{X^2}\,{}^V\Omega = (\gamma_X\Omega)^0 = {}^V(\gamma_X\Omega),$$
$$\gamma_{X^2}\,{}^I\Omega = (\gamma_X\Omega)^1 + (\gamma_X\pi)^0 = {}^I(\gamma_X\Omega),$$
$$\gamma_{X^2}\,{}^C\Omega = (\gamma_X\Omega)^2 + (\gamma_X\Omega)^1 + (\gamma_X\pi)^0 = {}^C(\gamma_X\Omega).$$

We shall now study the deformed lifts of exterior differentials of 1-forms $\omega = \omega_i \mathrm{d}x^i, i = 1, \ldots, r$. Using $[X^2, Y^2] = [X, Y]^2$ and linearity of mappings $X \to X^0$, $X \to X^1$, $X \to X^2$, from Theorems 3.20 and 3.21 we have

$$\begin{aligned}
2(\mathrm{d}^I\omega)(X^2, Y^2) &= X^2({}^I\omega(Y^2)) - Y^2({}^I\omega(X^2)) - {}^I\omega([X^2, Y^2]) \\
&= X^2((\omega(Y))^1 + (H(Y))^0) - Y^2((\omega(X))^1 + (H(X))^0) - {}^I\omega([X, Y]^2) \\
&= (X\omega(Y))^1 + (XH(Y))^0 - (Y\omega(X))^1 - (YH(X))^0 - (\omega([X, Y]))^1 - (H([X, Y]))^0 \\
&= (X\omega(Y) - Y\omega(X) - \omega([X, Y]))^1 + (XH(Y) - YH(X) - H([X, Y]))^1 \\
&= 2((\mathrm{d}\omega)(X, Y))^1 + 2((\mathrm{d}H)(X, Y))^0 = 2(\gamma_Y(\mathrm{d}\omega)(X))^1 + 2(\gamma_Y(\mathrm{d}H)(X))^0 \\
&= 2(\gamma_Y(\mathrm{d}\omega))^1(X^2) + 2(\gamma_Y(\mathrm{d}H))^0(X^2) = 2(\gamma_{Y^2}(\mathrm{d}\omega)^1)(X^2) + 2(\gamma_{Y^2}(\mathrm{d}H)^0)(X^2) \\
&= 2((\mathrm{d}\omega)^1 + (\mathrm{d}H)^0)(X^2, Y^2).
\end{aligned}$$

By similar devices, we have

$$2(\mathrm{d}^C\omega)(X^2, Y^2) = 2((\mathrm{d}\omega)^2 + (\mathrm{d}G)^1 + (\mathrm{d}H)^0)(X^2, Y^2).$$

Since the arbitrary tensor field Ω of type (0.2) in $T^2(M_r)$ is completely determined by its action on the lifts X^2, Y^2 (see [87, p. 324]), i.e. if $\Omega(X^2, Y^2) = \tilde{\Omega}(X^2, Y^2)$ for any X, Y, then $\Omega = \tilde{\Omega}$, and we have

Theorem 3.22 *Let ω, G and H be 1-forms in M_r. Then the exterior differentials of the deformed intermediate and complete lifts of ω to $T^2(M_r)$ satisfy the following formulas:*

$$\mathrm{d}^I\omega = (\mathrm{d}\omega)^1 + (\mathrm{d}H)^0,$$
$$\mathrm{d}^C\omega = (\mathrm{d}\omega)^2 + (\mathrm{d}G)^1 + (\mathrm{d}H)^0.$$

3.13 Problems of Lifts in Symplectic Geometry

A manifold M is *symplectic* if it possesses a nondegenerate 2-form ω which is closed (i.e. $d\omega = 0$). For any manifold M of dimension n, the cotangent bundle T^*M is a natural symplectic $2n$-manifold with symplectic 2-form $\tilde{\omega} = -dp = dx^i \wedge dp_i$, where $p = p_i dx^i$ is the Liouville form (basic 1-form) on T^*M.

Let now (M, ω) be a symplectic manifold and $TM = \bigcup_{P\in M} T_P(M)$ $(T^*M = \bigcup_{P\in M} T_P^*(M))$ its tangent (cotangent) bundle. Suppose that the base space M is covered by a system of coordinate neighborhoods (U, x^i), where x^i, $i = 1, \ldots, n$ are the local coordinates in the neighborhood U. We introduce a system of local induced coordinates

$$(x^i, x^{\bar{i}}) = (x^i, v^i)((x^i, \tilde{x}^{\bar{i}}) = (x^i, p_i)), \quad \bar{i} = n+1, \ldots, 2n$$

in the open set $\pi^{-1}(U) \subset TM$ $(\pi^{-1}(U) \subset T^*M)$. Then the symplectic isomorphisms $\omega^b : TM \to T^*M$ and $\omega^\sharp : T^*M \to TM$ are given by

$$\omega^b : x^I = (x^i, x^{\bar{i}}) = (x^i, v^i) \to \tilde{x}^K = (x^k, \tilde{x}^{\bar{k}}) = (x^k = \delta_i^k x^i, p_k = \omega_{ki} v^i)$$

and

$$\omega^\sharp : \tilde{x}^K = (x^k, \tilde{x}^{\bar{k}}) = (x^k, p_k) \to x^I = (x^i, x^{\bar{i}}) = (x^i = \delta_k^i x^k, v^i = \omega^{ik} p_k),$$

where $\omega^{ik}\omega_{kj} = \delta_j^i$, δ_j^i is the Kronecker symbol. The Jacobian matrices of ω^b and $\omega^\sharp$ are given respectively by

$$(\omega^b)_* = \tilde{A} = (\tilde{A}_I^K) = \begin{pmatrix} \tilde{A}_i^k & \tilde{A}_{\bar{i}}^k \\ \tilde{A}_i^{\bar{k}} & \tilde{A}_{\bar{i}}^{\bar{k}} \end{pmatrix} = \left(\frac{\partial \tilde{x}^K}{\partial x^I}\right) = \begin{pmatrix} \delta_i^k & 0 \\ v^s \partial_i \omega_{ks} & \omega_{ki} \end{pmatrix} \tag{3.59}$$

and

$$(\omega^\sharp)_* = A = (A_K^I) = \begin{pmatrix} A_k^i & A_{\bar{k}}^i \\ A_k^{\bar{i}} & A_{\bar{k}}^{\bar{i}} \end{pmatrix} = \left(\frac{\partial x^I}{\partial \tilde{x}^K}\right) = \begin{pmatrix} \delta_k^i & 0 \\ p_s \partial_k \omega^{is} & \omega^{ik} \end{pmatrix}. \tag{3.60}$$

Let f be any function on a symplectic manifold (M, ω). If ${}^C X_T$ is the complete lift of a vector field X from manifold M to its tangent bundle TM which is defined by

$${}^C X_T \, {}^C f = {}^C(Xf), \quad {}^C f = v^s \partial_s f,$$

then ${}^C X_T$ has the components [87, p. 15]

$${}^C X_T = \begin{pmatrix} X^i \\ v^s \partial_s X^i \end{pmatrix} \tag{3.61}$$

with respect to the coordinates $(x^i, x^{\bar{i}}) = (x^i, v^i)$.

Using (3.59) and (3.61) we have

$$(\omega^b)_* {}^C X_T = (\tilde{A}_I^K {}^C X_T^I) = \begin{pmatrix} \delta_i^k & 0 \\ v^s \partial_i \omega_{ks} & \omega_{ki} \end{pmatrix} \begin{pmatrix} X^i \\ v^s \partial_s X^i \end{pmatrix}$$

$$= \begin{pmatrix} X^k \\ X^i v^s \partial_i \omega_{ks} + \omega_{ki} v^s \partial_s X^i \end{pmatrix}$$

$$= \begin{pmatrix} X^k \\ v^s (X^i \partial_i \omega_{ks} + \omega_{is} \partial_k X^i + \omega_{ki} \partial_s X^i) - v^s \omega_{is} \partial_k X^i \end{pmatrix}$$

$$= \begin{pmatrix} X^k \\ v^s L_X \omega_{ks} - p_i \partial_k X^i \end{pmatrix}, \tag{3.62}$$

where L_X denotes the Lie derivative.

On other hand, the complete lift ${}^C X_{T^*}$ of a vector field X from manifold M to its cotangent bundle T^*M is defined by

$${}^C X_{T^*}(\gamma Z) = \gamma(L_X Y),$$

where γZ and $\gamma(L_X Y)$ are functions in T^*M with the local expressions

$$\gamma Z = p_i Z^i, \quad \gamma(L_X Z) - p_i [X, Z]^i$$

and the complete lift ${}^C X_{T^*}$ has the components [87, p. 236]

$${}^C X_{T^*} = \begin{pmatrix} X^k \\ -p_i \partial_k X^i \end{pmatrix}$$

with respect to the coordinates $(x^i, x^{\bar{i}}) = (x^i, p_i)$.

From (3.62) we obtain

$$(\omega^b)_* {}^C X_T = {}^C X_{T^*} + \begin{pmatrix} 0 \\ v^s L_X \omega_{ks} \end{pmatrix},$$

i.e. if $L_X \omega_{ks} = 0$, then $(\omega^b)_* {}^C X_T = {}^C X_{T^*}$. A *symplectic vector field* X is a vector field on (M, ω) which preserves the symplectic form, i.e. $L_X \omega = 0$. Thus we have

Theorem 3.23 *Let (M, ω) be a symplectic manifold, ${}^C X_T$ and ${}^C X_{T^*}$ the complete lifts of a vector field X to tangent bundle TM and cotangent bundle T^*M, respectively. If X is a symplectic vector field, then ${}^C X_T$ and ${}^C X_{T^*}$ are ω^b-related, i.e. $(\omega^b)_* {}^C X_T = {}^C X_{T^*}$.*

Since every Hamiltonian vector field X_H ($\iota_{X_H}\omega = \mathrm{d}H$) is a symplectic vector field ($L_{X_H}\omega = d\circ\iota_{X_H}\omega + \mathrm{d}\iota_{X_H}\circ\mathrm{d}\omega = \mathrm{d}^2H = 0$), from Theorem 3.23 we immediately have

Theorem 3.24 *If* X_H *is a Hamiltonian vector field, then* ${}^C(X_H)_T$ *and* ${}^C(X_H)_{T^*}$ *are* ω^b*-related.*

Let (M,ω) be a symplectic manifold of dimension $n = 2m$. It is well known that in the cotangent bundle T^*M there exists a closed 2-form $\tilde{\omega} = \mathrm{d}p = \mathrm{d}p_i \wedge \mathrm{d}x^i$, where $p = p_i\mathrm{d}x^i$, i.e. T^*M is a symplectic $4m$-manifold. If we write $\tilde{\omega} = \frac{1}{2}\tilde{\omega}_{KL}\mathrm{d}x^K \wedge \mathrm{d}x^L$, then we have

$$\tilde{\omega} = (\tilde{\omega}_{KL}) = \begin{pmatrix} 0 & \delta_k^l \\ -\delta_l^k & 0 \end{pmatrix}.$$

The complete lift ${}^C\omega_T$ of ω to the tangent bundle TM is a 2-form and has the components of the form [87, p. 38]

$${}^C\omega_T = \begin{pmatrix} v^s\partial_s\omega_{ij} & \omega_{ij} \\ \omega_{ij} & 0 \end{pmatrix} \tag{3.63}$$

with respect to the coordinates $(x^i, x^{\bar{i}}) = (x^i, v^i)$.

We now consider the symplectic isomorphism $\omega^\sharp : T^*M \to TM$. Using

$$(\mathrm{d}\omega)_{skl} = \tfrac{1}{3}(\partial_s\omega_{kl} + \partial_k\omega_{ls} + \partial_l\omega_{sk}) = 0, \quad \omega_{ij} = -\omega_{ji},$$
$$\omega^{ij} = -\omega^{ji}, \quad \omega^{is}\omega_{sj} = \delta^i_j, \tag{3.64}$$

from (3.60) and (3.63) we see that the pullback of ${}^C\omega$ by $\omega^\sharp$ is a 2-form $(\omega^\sharp)^*\,{}^C\omega_T$ on T^*M and has the components

$$\begin{aligned}
((\omega^\sharp)^*\,{}^C\omega_T)_{kl} &= A_k^I A_l^J({}^C\omega_T)_{IJ} = A_k^i A_l^j({}^C\omega_T)_{ij} + A_k^{\bar{i}} A_l^j({}^C\omega_T)_{\bar{i}j} + A_k^i A_l^{\bar{j}}({}^C\omega_T)_{i\bar{j}} \\
&= \delta_k^i\delta_l^j v^s\partial_s\omega_{ij} + p_s(\partial_k\omega^{is})\delta_l^j\omega_{ij} + \delta_k^i p_s(\partial_l\omega^{js})\omega_{ij} \\
&= v^s\partial_s\omega_{kl} + p_s((\partial_k\omega^{is})\omega_{il} - (\partial_l\omega^{sj})\omega_{kj}) \\
&= p_t\omega^{ts}\partial_s\omega_{kl} - p_s(\omega^{si}\partial_k\omega_{il} - \omega^{sj}\partial_l\omega_{jk}) \\
&= p_t\omega^{ts}(\partial_s\omega_{kl} - \partial_k\omega_{sl} + \partial_l\omega_{sk}) \\
&= 3p_t\omega^{ts}(\mathrm{d}\omega)_{skl} = 0,
\end{aligned}$$
$$((\omega^\sharp)^*\,{}^C\omega_T)_{k\bar{l}} = A_k^i A_{\bar{l}}^{\bar{j}}({}^C\omega_T)_{i\bar{j}} = \delta_k^i\omega^{jl}\omega_{ij} = \delta_k^i\delta_i^l = \delta_k^l,$$
$$((\omega^\sharp)^*\,{}^C\omega_T)_{\bar{k}l} = A_{\bar{k}}^{\bar{i}} A_l^j({}^C\omega_T)_{\bar{i}j} = \omega^{ik}\delta_l^j\omega_{ij} = -\delta_j^k\delta_l^j = -\delta_l^k,$$
$$((\omega^\sharp)^*\,{}^C\omega_T)_{\bar{k}\bar{l}} = 0$$

or

$$(\omega^\sharp)^* {}^C\omega_T = (((\omega^\sharp)^* {}^C\omega_T)_{KL}) = \begin{pmatrix} 0 & \delta^l_k \\ -\delta^k_l & 0 \end{pmatrix} .$$

From here it follows that the pullback $(\omega^\sharp)^* {}^C\omega_T$ coincides with the symplectic form $\tilde{\omega} = \mathrm{d}p = \mathrm{d}p_i \wedge \mathrm{d}x^i$. Thus we have

Theorem 3.25 *Let (M, ω) be a symplectic manifold. The natural symplectic structure $\mathrm{d}p = \mathrm{d}p_i \wedge \mathrm{d}x^i$ on the cotangent bundle T^*M is the pullback by $\omega^\sharp$ of the complete lift of ω to the tangent bundle TM, i.e.$(\omega^\sharp)^* {}^C\omega_T = dp$.*

A diffeomorphism between any two symplectic manifods $f : (M, \omega) \to (N, \omega')$ is called *symplectomorphism* if $f^*\omega' = \omega$, where f^* is the pullback of f. Since $\mathrm{d}\, {}^C\omega_T = {}^C(\mathrm{d}\omega)_T = 0$ [87, p. 25], from Theorem 3.25 we have that the symplectic isomorphism $\omega^\sharp : (T^*M, \mathrm{d}p) \to (TM, {}^C\omega_T)$ is a symplectomorphism.

Let now (M, ω) be a symplectic manifold with almost complex structure φ ($\varphi^2 = -I$). If the 2-form ω satisfies the purity condition $\omega(\varphi X, Y) = \omega(X, \varphi Y)$, i.e. $(\omega \circ \varphi)(X, Y) = -(\omega \circ \varphi)(Y, X)$, then the triple (M, ω, φ) is called *A-manifold* according to the terminology accepted in [40] (also, see [81, p. 31]). We call $\Omega(X, Y) = (\omega \circ \varphi)(X, Y) = \omega(\varphi Y, X)$ the twin 2-form associated with ω.

Let $\mathbb{C}$ be a complex algebra and $\overset{*}{\omega} = (\overset{*}{\omega}_{v_1 v_2})$, $v_1, v_2 = 1, \ldots, r$ be a complex tensor field of type (0,2) on the holomorphic (analytic) complex manifold $\mathfrak{X}_r(\mathbb{C})$. Then the real model of $\overset{*}{\omega}$ is a tensor field $\omega = (\omega_{j_1 j_2})$, $j_1, j_2 = 1, \ldots, 2r$ on M such that

$$\omega(\varphi X_1, X_2) = \omega(X_1, \varphi X_2)$$

for any vector fields X_1, X_2.

The Φ_φ-operator applied to a pure tensor field ω is defined by

$$\begin{aligned}(\Phi_\varphi \omega)(X, Y_1, Y_2) &= (\varphi X)(\omega(Y_1, Y_2)) - X(\omega(\varphi Y_1, Y_2)) \\ &\quad + \omega((L_{Y_1}\varphi)X, Y_2) + \omega(Y_1, (L_{Y_2}\varphi)X)\end{aligned}$$

and has the local expression

$$(\Phi_\varphi \omega)_{kij} = \varphi^m_k \partial_m \omega_{ij} - \partial_k (\omega \circ \varphi)_{ij} + \omega_{mj} \partial_i \varphi^m_k + \omega_{im} \partial_j \varphi^m_k , \tag{3.65}$$

where $\Phi_\varphi \omega$ is a tensor field of type (0,3), L_X is the Lie derivative with respect to X and

$$(\omega \circ \varphi)_{ij} = \varphi^m_i \, \omega_{mj} \ .$$

Let on M be given the integrable almost complex structure φ. For a complex tensor field $\overset{*}{\omega}$ of type (0,2) on $\mathfrak{X}_r(\mathbb{C})$ to be a $\mathbb{C}$-holomorphic tensor field it is necessary and sufficient that $\Phi_\varphi \omega = 0$. Let now M be a manifold with non-integrable almost complex structure φ. In this case, when $\Phi_\varphi \omega = 0$, ω is said to be almost holomorphic. If the symplectic 2-form ω of the A-manifold (M, ω, J) satisfies the almost holomorphicity condition $\Phi_\varphi \omega = 0$, then it is called an almost holomorphic symplectic 2-form. We call an A-manifold admitting such a 2-form an almost *holomorphic A-manifold.*

Let $\varphi = \varphi_j^i \, \partial_i \otimes dx^j$ be a tensor field of type (1,1) in $U \subset M$. The complete lift ${}^C\varphi_{TM}$ of φ to the tangent bundle is completely determined by ${}^C\varphi_{TM}({}^CX) = {}^C(\varphi(X))_{TM}$. In an analogous way, the complete lift ${}^C\varphi_{T^*M}$ of φ to the cotangent bundle is completely determined by ${}^C\varphi_{T^*M}({}^CX) = {}^C(\varphi(X))_{T^*M} + \gamma(L_X\varphi)$, where $\gamma(L_X\varphi)$ is a vertical vector field on T^*M with the components $\gamma(L_X\varphi) = \sum_{i=1}^n p_s (L_X\varphi)_i^s \partial_{\bar{i}}$. The complete lift of φ to the tangent and cotangent bundles are given respectively by [87]

$$
{}^C\varphi_{TM} = (({}^C\varphi_{TM})_J^I) = \begin{pmatrix} \varphi_j^i & 0 \\ v^s \partial_s \varphi_j^i & \varphi_j^i \end{pmatrix}
$$

and

$$
{}^C\varphi_{T^*M} = (({}^C\varphi_{T^*M})_J^I) = \begin{pmatrix} \varphi_j^i & 0 \\ p_s(\partial_j \varphi_i^s - \partial_i \varphi_j^s) & \varphi_i^j \end{pmatrix}
$$

with respect to the induced coordinates $(x^j, x^{\bar{j}}) = (x^j, v^j)$ and $(x^j, x^{\bar{j}}) = (x^j, p_j)$.

Using (3.59), (3.60), (3.64) and $\varphi_j^m \omega_{mk} = \varphi_k^m \omega_{jm}$ to transfer ${}^C\varphi_{TM}$ by $\omega^\sharp : T^*M \to TM$ we have

$$
(\omega^\sharp)^* {}^C\varphi_{TM} = ((\widetilde{\varphi}_{T^*M})_L^J) = (\tilde{A}_I^J A_L^K ({}^C\varphi_{TM})_K^I)
$$

or

$$
\begin{aligned}
(\widetilde{\varphi}_{T^*M})_l^j &= \varphi_l^j, \quad (\widetilde{\varphi}_{T^*M})_{\bar{l}}^j = 0, \quad (\widetilde{\varphi}_{T^*M})_{\bar{l}}^{\bar{j}} = \omega_{ji}\omega^{kl}\varphi_k^i = \varphi_j^l, \\
(\widetilde{\varphi}_{T^*M})_l^{\bar{j}} &= v^s(\partial_i \omega_{js})\varphi_l^i + \omega_{ji} v^s \partial_s \varphi_l^i + \omega_{ji} p_s (\partial_l \omega^{ks})\varphi_k^i \\
&= v^s\left((\Phi_\varphi\omega)_{ljs} + \partial_l(\omega \circ \varphi)_{js} - \omega_{is}\partial_j \varphi_l^i\right) + \omega_{ji} p_s(\partial_l \omega^{ks})\varphi_k^i \\
&= v^s(\Phi_\varphi\omega)_{ljs} - p_i \partial_j \varphi_l^i + v^s \partial_l\left(\varphi_j^m \omega_{ms}\right) + \omega_{ji} p_s(\partial_l \omega^{ks})\varphi_k^i \\
&= v^s(\Phi_\varphi\omega)_{ljs} - p_i \partial_j \varphi_l^i + v^s\left(\partial_l \varphi_j^m\right)\omega_{ms} + v^s \varphi_j^m(\partial_l \omega_{ms}) + \omega_{jm} p_s(\partial_l \omega^{ks})\varphi_k^m \\
&= v^s(\Phi_\varphi\omega)_{ljs} - p_i \partial_j \varphi_l^i + p_m \partial_l \varphi_j^m + v^s(\partial_l \omega_{ms})\varphi_j^m + \omega_{mk} p_s(\partial_l \omega^{ks})\varphi_j^m \\
&= v^s(\Phi_\varphi\omega)_{ljs} + p_m(\partial_l \varphi_j^m - \partial_j \varphi_l^m) + v^s(\partial_l \omega_{ms})\varphi_j^m - \omega^{ks} p_s(\partial_l \omega_{mk})\varphi_j^m \\
&= v^s(\Phi_\varphi\omega)_{ljs} + p_m(\partial_l \varphi_j^m - \partial_j \varphi_l^m) + v^s(\partial_l \omega_{ms})\varphi_j^m - v^k(\partial_l \omega_{mk})\varphi_j^m \\
&= v^s(\Phi_\varphi\omega)_{ljs} + p_m(\partial_l \varphi_j^m - \partial_j \varphi_l^m).
\end{aligned}
$$

Thus, if $\Phi_\varphi \omega = 0$, then the transfer $(\omega^\sharp)^* {}^C\varphi_{TM}$ of ${}^C\varphi_{TM}$ coincides with ${}^C\varphi_{T^*M}$. Thus we have

Theorem 3.26 *Let (M, ω, φ) be a symplectic A-manifold and $\omega^\sharp : T^*M \to TM$ be a symplectic isomorphism between cotangent and tangent bundles. If the symplectic A-manifold is an almost holomorphic ($\Phi_\varphi \omega = 0$), then the complete lift ${}^C\varphi_{T^*M}$ is a transfer of ${}^C\varphi_{TM}$ by $\omega^\sharp$, i.e. $(\omega^\sharp)^* {}^C\varphi_{TM} = {}^C\varphi_{T^*M}$.*

In the case of integrability of φ, the complete lifts ${}^C\varphi_{TM}$ and ${}^C\varphi_{T^*M}$ are complex structures on the tangent and cotangent bundles, respectively (see [87, p. 37, p. 256]), i.e. $(T^*M, {}^C\varphi_{T^*M})$ and $(TM, {}^C\varphi_{TM})$ are complex manifolds. Since $\tilde{A}^{-1} = A$ (see (3.59), (3.60)), the condition $(\omega^\sharp)^* {}^C\varphi_{TM} = (\tilde{A}_I^J A_L^K ({}^C\varphi_{TM})_K^I) = (({}^C\varphi_{T^*M})_L^J) = {}^C\varphi_{T^*M}$, in Theorem 3.26 can be written in the following form

$$ {}^C\varphi_{TM} \circ (\omega^\sharp)_* = (\omega^\sharp)_* \circ {}^C\varphi_{T^*M}, $$

where $(\omega^\sharp)_* = (A_J^I)$. From here it is clear that the mapping $\omega^\sharp : T^*M \to TM$ is holomorphic. Thus we have

Theorem 3.27 *Let (M, ω, φ) be a holomorphic symplectic A-manifold. If φ is an integrable almost complex structure, then the symplectic isomorphism $\omega^\sharp$ (or ω^b) is a holomorphic mapping.*

On the other hand, from (3.65) we obtain

$$ \begin{aligned} (\Phi_\varphi \omega)_{kij} &= \varphi_k^m \partial_m \omega_{ij} - \partial_k (\omega \circ \varphi)_{ij} + \omega_{mj} \partial_i \varphi_k^m + \omega_{im} \partial_j \varphi_k^m \\ &= \varphi_k^m (\partial_m \omega_{ij} - \partial_i \omega_{mj} - \partial_j \omega_{im}) + (\partial_i \omega_{mj}) \varphi_k^m + (\partial_j \omega_{im}) \varphi_k^m \\ &+ \omega_{mj} \partial_i \varphi_k^m + \omega_{im} \partial_j \varphi_k^m - \partial_k \big(\varphi_i^m \omega_{mj} \big) = \varphi_k^m (\partial_m \omega_{ij} + \partial_i \omega_{jm} + \partial_j \omega_{mi}) \\ &+ \partial_i (\varphi_k^m \omega_{mj}) + \partial_j (\varphi_k^m \omega_{im}) - \partial_k \big(\varphi_i^m \omega_{mj} \big) = 3 \varphi_k^m (\mathrm{d}\omega)_{mij} \\ &+ \partial_i \Omega_{kj} + \partial_j (\varphi_i^m \omega_{mk}) - \partial_k \Omega_{ij} = 3 \varphi_k^m (\mathrm{d}\omega)_{mij} + \partial_i \Omega_{kj} + \partial_j \Omega_{ik} + \partial_k \Omega_{ji} \\ &= 3 (\varphi_k^m (\mathrm{d}\omega)_{mij} + (\mathrm{d}\Omega)_{ikj}), \end{aligned} $$

which on symplectic A-manifold ($\mathrm{d}\omega = 0$) has the form

$$ (\Phi_\varphi \omega)(X, Y_1, Y_2) = 3(\mathrm{d}\Omega)(Y_1, X, Y_2), $$

where $\Omega = \omega \circ \varphi$ is the twin 2-form. Thus we have

Theorem 3.28 *A symplectic A-manifold (M, ω, φ) is holomorfic if and only if the twin 2-form $\Omega = \omega \circ \varphi$ is closed.*

From Theorems 3.27 and 3.28 we have

Theorem 3.29 *If* $\Omega = \omega \circ \varphi$ *is a closed twin 2-form on the A-manifold* (M, ω, φ), *then* ${}^{C}\varphi_{T^*M}$ *is a transfer of* ${}^{C}\varphi_{TM}$ *by the symplectic isomorphism* $\omega^{\sharp} : T^*M \to TM$.

3.14 Anti-Kähler Manifolds and Musical Isomorphisms

In pseudo-Riemannian geometry the *musical (bemolle and diesis) isomorphism* is an isomorphism between the tangent and cotangent bundles. Let (M, g) be a smooth pseudo-Riemannian manifold of dimension n. A very important feature of any pseudo-Riemannian metric g is that it provides the musical isomorphisms $g^{b} : TM \to T^*M$ and $g^{\sharp} : T^*M \to TM$ between the tangent and cotangent bundles. The musical isomorphisms g^{b} and $g^{\sharp}$ are expressed by

$$g^{b} : x^{I} = (x^{i}, x^{\bar{i}}) = (x^{i}, y^{i}) \to \tilde{x}^{K} = (x^{k}, \tilde{x}^{\bar{k}}) = (\delta_{i}^{k} x^{i}, p_{k} = g_{ki} y^{i})$$

and

$$g^{\sharp} : \tilde{x}^{K} = (x^{k}, \tilde{x}^{\bar{k}}) = (x^{k}, p_{k}) \to x^{I} = (x^{i}, x^{\bar{i}}) = (\delta_{k}^{i} x^{k}, y^{i} = g^{ik} p_{k})$$

with respect to the local coordinates in TM and T^*M respectively. The Jacobian matrices of g^{b} and $g^{\sharp}$ are given by

$$(g_{*}^{b}) = (\tilde{A}_{I}^{K}) = \left(\frac{\partial \tilde{x}^{K}}{\partial x^{I}}\right) = \begin{pmatrix} \delta_{i}^{k} & 0 \\ y^{s} \partial_{i} g_{ks} & g_{ki} \end{pmatrix} \tag{3.66}$$

and

$$(g_{*}^{\sharp}) = (A_{K}^{I}) = \left(\frac{\partial x^{I}}{\partial \tilde{x}^{K}}\right) = \begin{pmatrix} \delta_{k}^{i} & 0 \\ p_{s} \partial_{k} g^{is} & g^{ik} \end{pmatrix} \tag{3.67}$$

respectively, where δ is the Kronecker delta.

Let $X = X^{i} \partial_{i}$ be the local expression in $U \subset M$ of a vector field $X \in \Im_{0}^{1}(M)$. Then the complete lift ${}^{C}X_{T}$ of X to the tangent bundle TM is given by

$${}^{C}X_{T} = X^{i} \partial_{i} + y^{s} \partial_{s} X^{i} \partial_{\bar{i}} \tag{3.68}$$

with respect to the natural frame $\{\partial_{i}, \partial_{\bar{i}}\}$.

Using (3.66) and (3.68), we have

$$g_*^b\,{}^C X_T = \begin{pmatrix} \delta_i^k & 0 \\ y^s \frac{\partial g_{ks}}{\partial x^i} & g_{ki} \end{pmatrix} \begin{pmatrix} X^i \\ y^s \partial_s X^i \end{pmatrix} = \begin{pmatrix} X^k \\ X^i y^s \partial_i g_{ks} + g_{ki} y^s \partial_s X^i \end{pmatrix}$$
$$= \begin{pmatrix} X^k \\ y^s((L_X g)_{sk} - (\partial_k X^i) g_{is} - (\partial_s X^i) g_{ki}) + g_{ik} y^s \partial_s X^i \end{pmatrix}$$
$$= \begin{pmatrix} X^k \\ y^s (L_X g)_{sk} - p_i (\partial_k X^i) \end{pmatrix}, \tag{3.69}$$

where L_X is the Lie derivative of g with respect to the vector field X:

$$(L_X g)_{sk} = X^i \partial_l g_{sk} + \left(\partial_s X^i\right) g_{ik} + \left(\partial_k X^i\right) g_{si}.$$

In a manifold (M, g), a vector field X is called a *Killing vector field* if $L_X g = 0$. It is well known that the complete lift ${}^C X_{T^*}$ of X to the cotangent bundle T^*M is given by

$${}^C X_{T^*} = X^k \partial_k - p_s \partial_k X^s \partial_{\overline{k}}.$$

From (3.69) we find

$$g_*^b\,{}^C X_T = {}^C X_{T^*} + \gamma(L_X g),$$

where $\gamma(L_X g)$ is defined by

$$\gamma(L_X g) = \begin{pmatrix} 0 \\ y^s (L_X g)_{sk} \end{pmatrix}$$

Thus we have

Theorem 3.30 *Let (M, g) be a pseudo-Riemannian manifold, and let ${}^C X_T$ and ${}^C X_{T^*}$ be the complete lifts of a vector field X to the tangent and cotangent bundles, respectively. Then the differential (pushforward) of ${}^C X_T$ by g^b coincides with ${}^C X_{T^*}$, i.e. $g_*^b\,{}^C X_T = {}^C X_{T^*}$ if and only if X is a Killing vector field.*

Let X and Y be Killing vector fields on M. Then we have

$$L_{[X,Y]} g = [L_X, L_Y] g = L_X \circ L_Y g - L_Y \circ L_X g = 0,$$

i.e. $[X, Y]$ is a Killing vector field. Since ${}^C[X, Y]_T = [{}^C X_T, {}^C Y_T]$ and ${}^C[X, Y]_{T^*} = [{}^C X_{T^*}, {}^C Y_{T^*}]$, from Theorem 3.30 we have

Theorem 3.31 *If X and Y are Killing vector fields on M, then*

$$g_*^b [{}^C X_T, {}^C Y_T] = [{}^C X_{T^*}, {}^C Y_{T^*}],$$

where g^b_* is the differential (pushforward) of musical isomorphism g^b.

Let now (M, g, φ) be an anti-Kähler manifold, where φ denote its almost complex structure.

If $\varphi = \varphi^i_j\, \partial_i \otimes \mathrm{d}x^j$ is the local expression in $U \subset M$ of an almost complex strucure φ, then it is well known that the complete lift ${}^C\varphi_T$ of φ to the tangent bundle TM is given by

$$ {}^C\varphi_T = ({}^C\varphi^I_J) = \begin{pmatrix} \varphi^i_j & 0 \\ y^s \partial_s \varphi^i_j & \varphi^i_j \end{pmatrix} \tag{3.70}$$

with respect to the induced coordinates $(x^i, x^{\bar{i}}) = (x^i, y^i)$ in TM. It is also well known that ${}^C\varphi_T$ defines an almost complex structure on TM, if and only if so does φ on M.

Using (3.66), (3.67) and (3.70), we have

$$\begin{aligned} g^b_*\,{}^C\varphi_T &= (\widetilde{\varphi}^J_L) = (A^J_I \widetilde{A}^K_L\, {}^C\varphi^I_K) \\ &= \begin{pmatrix} \varphi^j_l & 0 \\ y^s(\partial_i g_{js})\varphi^i_l + g_{ji} y^s \partial_s \varphi^i_l + g_{ji} p_s (\partial_l g^{ks})\varphi^i_k & g_{ji} g^{kl} \varphi^i_k \end{pmatrix}. \end{aligned} \tag{3.71}$$

Since $g = (g_{ij})$ and $g^{-1} = (g^{ij})$ are pure tensor fields with respect to φ, we find

$$ g_{ji} g^{kl} \varphi^i_k = g_{ji} g^{ik} \varphi^l_k = \delta^k_j \varphi^l_k = \varphi^l_j \tag{3.72}$$

and

$$\begin{aligned} & y^s(\partial_i g_{js})\varphi^i_l + g_{ji} y^s \partial_s \varphi^i_l + g_{ji} p_s (\partial_l g^{ks})\varphi^i_k \\ &= y^s\left(\Phi_l g_{js} + \partial_l (g \circ \varphi)_{js} - g_{is}\partial_j \varphi^i_l\right) + g_{ji} p_s(\partial_l g^{ks})\varphi^i_k \\ &= y^s \Phi_l g_{sj} + y^s \partial_l (g \circ \varphi)_{js} - p_i \partial_j \varphi^i_l + g_{ji} p_s (\partial_l g^{ks})\varphi^i_k \\ &= y^s \Phi_l g_{sj} - p_i \partial_j \varphi^i_l + y^s \partial_l (g \circ \varphi)_{js} + g_{ji} p_s (\partial_l g^{ks})\varphi^i_k \\ &= y^s \Phi_l g_{sj} - p_i \partial_j \varphi^i_l + y^s \partial_l \left(g_{sm}\varphi^m_j\right) + g_{ji} p_s (\partial_l g^{ks})\varphi^i_k \\ &= y^s \Phi_l g_{sj} - p_i \partial_j \varphi^i_l + y^s (\partial_l g_{sm})\varphi^m_j + y^s\left(\partial_l \varphi^m_j\right) g_{sm} + g_{jm} p_s (\partial_l g^{ks})\varphi^m_k \\ &= y^s \Phi_l g_{sj} - p_i \partial_j \varphi^i_l + y^s (\partial_l g_{sm})\varphi^m_j + y^s\left(\partial_l \varphi^m_j\right) g_{sm} + g_{mk} p_s (\partial_l g^{ks})\varphi^m_j \\ &= y^s \Phi_l g_{sj} - p_i \partial_j \varphi^i_l + y^s (\partial_l g_{sm})\varphi^m_j + y^s\left(\partial_l \varphi^m_j\right) g_{sm} - g^{ks} p_s (\partial_l g_{mk})\varphi^m_j \\ &= y^s \Phi_l g_{sj} - p_i \partial_j \varphi^i_l + y^s (\partial_l g_{sm})\varphi^m_j + p_m\left(\partial_l \varphi^m_j\right) - y^k (\partial_l g_{mk})\varphi^m_j \\ &= y^s \Phi_l g_{sj} + p_s(\partial_l \varphi^s_j - \partial_j \varphi^s_l), \end{aligned} \tag{3.73}$$

where

$$\Phi_k g_{ij} = \varphi_k^m \partial_m g_{ij} - \partial_k (g \circ \varphi)_{ij} + g_{mj} \partial_i \varphi_k^m + g_{im} \partial_j \varphi_k^m .$$

Substituting (3.72) and (3.73) into (3.71), we obtain

$$g_*^b \, {}^C\varphi_T = \begin{pmatrix} \varphi_l^j & 0 \\ y^s \Phi_l g_{sj} + p_s(\partial_l \varphi_j^s - \partial_j \varphi_l^s) & \varphi_j^l \end{pmatrix}.$$

It is well known that the complete lift ${}^C\varphi_{T^*}$ of $\varphi \in \Im_1^1(M)$ to the cotangent bundle T^*M is given by

$${}^C\varphi_{T^*} = \begin{pmatrix} \varphi_l^j & 0 \\ p_s(\partial_l \varphi_j^s - \partial_j \varphi_l^s) & \varphi_j^l \end{pmatrix}$$

with respect to the induced cordinates in T^*M. Thus we obtain

$$g_*^b \, {}^C\varphi_T = {}^C\varphi_{T^*} + \gamma(\Phi_\varphi g),$$

where

$$\gamma(\Phi_\varphi g) = \begin{pmatrix} 0 & 0 \\ y^s \Phi_l g_{sj} & 0 \end{pmatrix}.$$

From here, we have

Theorem 3.32 *Let (M, g, φ) be an almost anti-Hermitian manifold, and let ${}^C\varphi_T$ and ${}^C\varphi_{T^*}$ be the complete lifts of an almost complex structure φ to the tangent and cotangent bundles, respectively. Then the differential of ${}^C\varphi_T$ by g^b coincides with ${}^C\varphi_{T^*}$, i.e. $g_*^b \, {}^C\varphi_T = {}^C\varphi_{T^*}$ if and only if (M, g, φ) is an anti-Kähler $(\Phi_\varphi g = 0)$ manifold.*

References

1. Bejan, C., Gul, I. Sasaki metric on the tangent bundle of a Weyl manifold. Publ. Inst. Math. (Beograd) (N.S.) 103 (117) (2018), 25–32.
2. Bejan, C.L., Druţă-Romaniuc, S.L.: Harmonic functions and quadratic harmonic morphisms on Walker spaces. Turk. J. Math. **40**(5), 1004–1019 (2016)
3. Bejan, C.L., Druţă-Romaniuc, S.L.: Structures which are harmonic with respect to Walker metrics. Mediterr. J. Math. **12**(2), 481–496 (2015)
4. Bejan, C.L., Druţă-Romaniuc, S.L.: Walker manifolds and Killing magnetic curves. Dif. Geom. Appl. **35**, 106–116 (2014)
5. Bejan, C.L., Crasmareanu, M.: Weakly-symmetry of the Sasakian lifts on tangent bundles. **83**(1–2), 63–69 (2013)
6. Bonome, A., Castro, R., Hervella, L.M., Matsushita, Y.: Construction of Norden structures on neutral 4-manifolds. JP J. Geom. Topol. **5**(2), 121–140 (2005)
7. Borowiec, A., Francaviglia, M., Volovich, I.: Anti-Kählerian manifolds. Dif. Geom. Appl. **12**(3), 281–289 (2000)
8. Cakan, R., Akbulut, K., Salimov, A.: Musical isomorphisms and problems of lifts. Chinese Ann. Math. Ser. B **37**(3), 323–330 (2016)
9. Cengiz, N., Salimov, A.A.: Complete lifts of derivations to tensor bundles. Bol. Soc. Mat. Mexicana. **8**(3), 75–82 (2002)
10. Cruceanu, V., Fortuny, P., Gadea, P.M.: A survey on paracomplex geometry. Rocky Mountain J. Math. **26**(1), 83–115 (1996)
11. Davidov, J., Díaz-Ramos, J.C., García-Río, E., Matsushita, Y., Muškarov, O., Vázquez-Lorenzo, R.: Almost Kähler Walker 4-manifolds. J. Geom. Phys. **57**, 1075–1088 (2007)
12. Davidov, J., Díaz-Ramos, J.C., García-Río, E., Matsushita, Y., Muškarov, O., Vázquez-Lorenzo, R.: Hermitian-Walker 4-manifolds. J. Geom. Phys. **58**, 307–323 (2008)
13. Dragomir, S., Francaviglia, M.: On Norden metrics which are locally conformal to anti-Kählerian metrics. Acta Appl. Math. **60**(2), 115–135 (2000)
14. Druţă-Romaniuc, S.L. General natural α-structures parallel with respect to the Schouten-Van Kampen connection on the tangent bundle. Mediterr. J. Math. 19 (4) (2022), Paper No. 195, 21 pp.
15. Druţă-Romaniuc, S.L.: General natural Riemannian almost product and para-Hermitian structures on tangent bundles. Taiwanese J. Math. **16**(2), 497–510 (2012)
16. Druţă-Romaniuc, S.L.: Classes of general natural almost anti-Hermitian structures on the cotangent bundles. Mediterr. J. Math. **8**(2), 161–179 (2011)
17. Etayo, F., deFrancisco, A., Santamaría, R. There are no genuine Kähler-Codazzi manifolds. Int. J. Geom. Methods Mod. Phys. 17 (2020), no. 3, 2050044, 12 pp.

A. Salimov, *Applications of Holomorphic Functions in Geometry*, Frontiers in Mathematics, https://doi.org/10.1007/978-981-99-1296-4

18. Etayo, F., Santamaría, R.: metric manifolds. Publ. Math. Debrecen. **57**(3–4), 435–444 (2000)
19. Gadea, P.M., Grifone, J., Munoz Masque, J.: Manifolds modelled over free modules over the double numbers. Acta Math. Hungar. **100**(3), 187–203 (2003)
20. Ganchev, G.T., Borisov, A.V.: Note on the almost complex manifolds with Norden metric. Compt. Rend. Acad. Bulg. Sci. **39**, 31–34 (1986)
21. García-Río, E., Matsushita, Y.: Isotropic Kähler structures on Engel 4-manifolds. J. Geom. Phys. **33**, 288–294 (2000)
22. Gribachev, K., Mekerov, D., Djelepov, G.: Generalized B-manifolds. C. R. Acad. Bulgare Sci. **38**(3), 299–302 (1985)
23. Gribachev, K., Mekerov, D., Djelepov, G.: On the geometry of almost B-manifolds. C. R. Acad. Bulgare Sci. **38**(5), 563–566 (1985)
24. Gudmundsson, S., Kappos, E.: On the Geometry of the Tangent Bundles. Expo. Math. **20**(1), 1–41 (2002)
25. Iscan, M., Salimov, A.A. On Kahler-Norden manifolds. Proc. Indian Acad. Sci. (Math. Sci.) 119 (2009), no.1, 71–80.
26. Kobayashi, S., Nomizu, K.: Foundations of differential geometry, vol. II. Interscience Publishers, New York-London-Sydney (1969)
27. Kruchkovich, G.I.: Conditions for the integrability of a regular hypercomplex structure on a manifold. Ukrain. Geometr. Sb. **9**, 67–75 (1970)
28. Kruchkovich, G.I.: Hypercomplex structures on manifolds. I. Trudy Sem. Vektor. Tenzor. Anal. **16**, 174–201 (1972)
29. Kruchkovich, G.I.: Hypercomplex structures on manifolds. II. Trudy Sem. Vektor. Tenzor. Anal. **17**, 218–227 (1974)
30. Ledger, A.J., Yano, K.: Almost complex structures on tensor bundles. J. Dif. Geom. **1**, 355–368 (1967)
31. Magden, A., Salimov, A.A.: Complete lifts of tensor fields on a pure cross-section in the tensor bundle. J. Geom. **93**(1–2), 128–138 (2009)
32. Manev, M., Mekerov, D.: On Lie groups as quasi-Kähler manifolds with Killing Norden metric. Adv. Geom. **8**(3), 343–352 (2008)
33. Matsushita, Y., Law, P.: Hitchin-Thorpe-type inequalities for pseudo-Riemannian 4-manifolds of metric signature. Geom. Ded. **87**, 65–89 (2001)
34. Matsushita, Y.: Four-dimensional Walker metrics and symplectic structure. J. Geom. Phys. **52**, 89–99 (2004)
35. Matsushita, Y.: Walker 4-manifolds with proper almost complex structure. J. Geom. Phys. **55**, 385–398 (2005)
36. Matsushita, Y. Counterexamples of compact type to the Goldberg conjecture and various version of the conjecture. Proceedings of The 8th International Workshop on Complex Structures and Vector Fields, Sofia, Bulgaria, August 20 - 26, (2004), ed. S. Dimiev and K. Sekigawa, World Scientific (2007).
37. Matsushita, Y., Haze, S., Law, P.R.: Almost Kähler-Einstein structure on 8-dimensional walker manifolds. Monatsh. Math. **150**, 41–48 (2007)
38. Mekerov, D.: A connection with skew symmetric torsion and Kähler curvature tensor on quasi-Kähler manifolds with Norden metric. C. R. Acad. Bulgare Sci. **61**(10), 1249–1256 (2008)
39. Mekerov, D.: Connection with parallel totally skew-symmetric torsion on almost complex manifolds with Norden metric. C. R. Acad. Bulgare Sci. **62**(12), 1501–1508 (2009)
40. Norden, A.P.: On a class of four-dimensional A-spaces. Izv. Vyssh. Uchebn. Zaved. Matematika. **17**(4), 145–157 (1960)
41. O'Neill, B.: Semi-Riemannian geometry. Academic Press, New York (1983)

42. Oproiu, V., Papaghiuc, N.: Some classes of almost anti-Hermitian structures on the tangent bundle. Mediterr. J. Math. **1**(3), 269–282 (2004)
43. Oproiu, V., Papaghiuc, N.: An anti-Kählerian Einstein structure on the tangent bundle of a space form. Colloq. Math. **103**(1), 41–46 (2005)
44. Popovici, I.: Contributions à l'étude des espaces à connexion constante. Rev. Roumaine Math. Pures Appl. **23**(8), 1211–1225 (1978)
45. Salimov, A.A. Almost analyticity of a Riemannian metric and integrability of a structure.Trudy Geom. Sem. Kazan. Univ. 15 (1983), 72–78.
46. Salimov, A.A. Quasiholomorphic mapping and a tensor bundle.Translation in Soviet Math. (Iz. VUZ) 33 (1989), no. 12, 89–92.
47. Salimov, A.A. Almost -holomorphic tensors and their properties. Translation in Russian Acad. Sci. Dokl. Math. 45 (1992), no. 3, 602–605.
48. Salimov, A.A. A new method in the theory of liftings of tensor fields in a tensor bundle. Translation in Russian Math. (Iz. VUZ) 38 (1994), no. 3, 67–73.
49. Salimov, A.A. Generalized Yano-Ako operator and the complete lift of tensor fields.Tensor (N.S.) 55 (1994), no. 2, 142–146.
50. Salimov, A.A. Lifts of poly-affinor structures on pure sections of a tensor bundle. Translation in Russian Math. (Iz. VUZ) 40 (1996), no. 10, 52–59.
51. Salimov, A.A.: Non-existence of Para-Kahler-Norden warped metrics. Int. J. Geom. Methods Mod. Phys. **6**(7), 1097–1102 (2009)
52. Salimov, A.: On operators associated with tensor fields. J. Geom. **99**(1–2), 107–145 (2010)
53. Salimov, A.A.: A note on the Goldberg conjecture of Walker manifolds. Int. J. Geom. Methods Mod. Phys. **8**(5), 925–928 (2011)
54. Salimov, A. Tensor operators and their applications. Mathematics Research Developments. Nova Science Publishers, Inc., New York, 2013. xii+186 pp.
55. Salimov, A.: On anti-Hermitian metric connections. C. R. Math. Acad. Sci. Paris. **352**(9), 731–735 (2014)
56. Salimov, A.: On structure-preserving connections. Period. Math. Hungar. **77**(1), 69–76 (2018)
57. Salimov, A.A., Agca, F.: On para-Nordenian structures. Ann. Polon. Math. **99**(2), 193–200 (2010)
58. Salimov, A., Aslanci, S., Jabrailzade, F.: Dual-holomorphic functions and problems of lifts. Chinese Ann. Math. Ser. B **43**(2), 223–232 (2022)
59. Salimov, A., Behboudi Asl, M., Kazimova, S.: Problems of lifts in symplectic geometry. Chinese Ann. Math. Ser. B **40**(3), 321–330 (2019)
60. Salimov, A.A., Cengiz, N., Behboudi Asl, M.: On holomorphic hypercomplex connections. Adv. Appl. Clifford Algebr. **23**(1), 179–207 (2013)
61. Salimov, A., Gezer, A.: On the geometry of the (1,1)-tensor bundle with Sasaki type metric. Chin. Ann. Math. Ser. B **32**(3), 369–386 (2011)
62. Salimov, A.A., Gezer, A., Akbulut, K.: Geodesics of Sasakian metrics on tensor bundles. Mediterr. J. Math. **6**(2), 135–147 (2009)
63. Salimov, A., Iscan, M. Some properties of Norden-Walker metrics.Kodai Math. J. 33 (2010), no.2, 283–293.
64. Salimov, A.A., Iscan, M.: On Norden-Walker 4-manifolds. Note Mat. **30**, 111–128 (2010)
65. Salimov, A., Iscan, M., Akbulut, K.: Some remarks concerning hyperholomorphic B-manifolds. Chin. Ann. Math. Ser. B **29**(6), 631–640 (2008)
66. Salimov, A.A., Iscan, M., Akbulut, K. Notes on para-Norden-Walker 4-manifolds.Int. J. Geom. Methods Mod. Phys. 7 (2010), no.8, 1331–1347.
67. Salimov, A.A., Iscan, M., Etayo, F. Paraholomorphic B-manifold and its properties.Topology Appl. 154 (2007), no. 4, 925–933.

68. Salimov, A.A., Magden, A.: Complete lifts of tensor fields on a pure cross-section in the tensor bundle. Note Mat. **18**(1), 27–37 (1998)
69. Sasaki, S.: On the Differantial geometry of tangent bundles of Riemannian manifols. Tohoku Math. J. **10**, 338–358 (1958)
70. Sato, I. Almost analytic tensor fields in almost complex manifolds. Tensor (N.S.) 17 (1966), 105–119.
71. Scheffers, G. Generalization of the foundations of ordinary complex functions. I, II. (Verallgemeinerung der Grundlagen der gewöhnlich complexen Functionen. I. II.) (German) Leipz. Ber. XLV. (1893), 828–848.
72. Sekigawa, K.: On some 4-dimensional compact Einstein almost Kähler manifolds. Math. Ann. **271**(3), 333–337 (1985)
73. Shirokov, A.P. On a property of covariantly constant affinors. Dokl. Akad. Nauk SSSR (N.S.) 102 (1955), 461–464.
74. Shirokov, A.P. On the question of pure tensors and invariant subspaces in manifolds with almost algebraic structure. Kazan. Gos. Univ. Učen. Zap. 126 (1966), kn. 1, 81–89.
75. Shirokov, A.P. Spaces over algebras and their applications. Geometry, 7. J. Math. Sci. (New York) 108 (2002), no. 2, 232–248.
76. Sultanova, T., Salimov, A. On holomorphic metrics of 2-jet bundles. Mediterr. J. Math. 19 (2022), no. 1, Paper No. 29, 12 pp.
77. Tachibana, S.: Analytic tensor and its generalization. Tohoku Math. J. **12**, 208–221 (1960)
78. Talantova, N.V., Shirokov, A.P.: A remark on a certain metric in the tangent bundle. Izv. Vysš. Učebn. Zaved. Matematika **157**(6), 143–146 (1975)
79. Vishnevskii, V.V. On the complex structure of B-spaces.Kazan. Gos. Univ. Uchen. Zap. 123 (1963), kn. 1, 24–48.
80. Vishnevskii, V.V.: A certain class of spaces over plural algebras. Izv. Vysš. Učebn. Zaved. Matematika **81**(2), 14–22 (1969)
81. Vishnevskii, V.V., Shirokov, A.P., Shurygin, V.V.: Spaces over algebras. Kazanskii Gosudarstvennii Universitet, Kazan (1985)
82. Vishnevskii, V.V. Integrable affinor structures and their plural interpretations. Geometry, 7. J. Math. Sci. (New York) 108 (2002), no. 2, 151–187.
83. Vranceanu, G. Spazi a connessione affine e le algebre di numeri ipercomplessi. Ann. Scuola Norm. Sup. Pisa (3) 12 (1958), 5–20.
84. Walker, A.G.: Canonical form for a Riemannian space with a paralel field of null planes. Quart. J. Math. Oxford **1**(2), 69–79 (1950)
85. Yano, K., Kobayashi, Sh. Prolongations of tensor fields and connections to tangent bundles. I. General theory. J. Math. Soc. Japan 18 (1966), 194–210.
86. Yano, K., Ako, M.: On certain operators associated with tensor fields. Kodai Math. Sem. Rep. **20**, 414–436 (1968)
87. Yano, K., Ishihara, S. Tangent and cotangent bundles: differential geometry. Pure and Applied Mathematics, No. 16. Marcel Dekker, Inc., New York, 1973.

Index

A. Salimov, *Applications of Holomorphic Functions in Geometry*, Frontiers in Mathematics, https://doi.org/10.1007/978-981-99-1296-4